Mentals 4

Alan McSeveny Rachel McSeveny Diane McSeveny-Foster

Pearson Australia
(a division of Pearson Australia Group Pty Ltd)
459–471 Church St, Level 1, Building B, Richmond, Victoria, 3121
PO Box 23360, Melbourne, Victoria 8012
www.pearson.com.au

First published 2024 by Pearson Australia
2028 2027 2026 2025
10 9 8 7 6 5 4 3 2 1

Publishers: Sophie Matta and Kerry Nagle
Project Manager: Michelle Thomas
Production Editor: Laura Rentsch
Editor: Rachel Elliott
Designer: Anne Donald
Proofreader: Ann M. Philpott
Rights & Permissions Editor: Alice McBroom
Cover art: Michael Barter
Illustrator: Michael Barter
Desktop Operator: Jit-Pin Chong
Printed in Australia by Pegasus Media + Logistics

ISBN 978 0 655 708 841
Pearson Australia Group Pty Ltd ABN 40 004 245 943

Acknowledgement of Country
Pearson respects and honours Aboriginal and Torres Strait Islander Elders past, present and future. We acknowledge the stories, traditions and living cultures of the Traditional Custodians of the lands on which our company is located and where we conduct our business. Pearson is committed to honouring Australian Aboriginal and Torres Strait Islander peoples' unique cultural and spiritual relationships to the land, waters and seas and their rich contribution to society.

Aboriginal and Torres Strait Islander peoples are advised that this text may contain images, voices and names of deceased persons.

Introduction

Using the Mentals Books

This book is used most effectively when it aligns with the suggested program in the Student Book contents. Each unit of the Mentals Book is programmed to review Student Book content for the previous two weeks (based on the Suggested Program in the Teacher's Book). For example, Unit 15 of the Mentals Book can be set as homework to review weeks 13 and 14 of the Student Book while week 15 is being taught.

Mixed-topic questions

The units present questions in a mixed-topic format to encourage thorough understanding and continuous review.

Motivation

- There are two lizards hidden on each page for students to find.
- The header allows students to record their score.

Presentation

- Number facts are reinforced to encourage instant recall.
- Essential skills are explained.
- The Arithmetic card (page 5) is a useful teaching tool for practising basic number skills.
- ID cards (pages 6, 7 and 8) review the mathematical terms students need to learn.
- Examples of measurements (page 9) are provided so that students can learn important facts and estimate measurements effectively.

Graded questions

- Column 1: easier
- Column 2 and 3: harder
- Column 4: Extension and Challenge

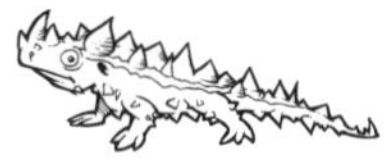

Extra activities

- Problem-solving **strategies** are introduced in a carefully planned sequence throughout the series.

- Important concepts from **Number and algebra** and **Measurement and geometry** are explored.

- **Measurement** concepts and activities are introduced and investigated.

- **Statistics and probability** concepts (Data and chance) are presented for revision and extension.

- A **tables** program for each of the four operations is included.
- It is important for students to learn addition and multiplication tables by heart.

4 Contents

Unit activities

Unit	Content	Extra Activity	Unit	Content	Extra Activity
1:1/2 **1:3/4**	+ 2,+ 3,+ 4 Personal measures	+ tables Measure	**20:1/2** **20:3/4**	Language Problem solving	ID card C Strategy time
2:1/2 **2:3/4**	− 2, − 4 + 5, + 10, + 6	− tables + tables	**21:1/2** **21:3/4**	− 6, − 7 × 6, × 8	− tables × tables
3:1/2 **3:3/4**	Language + 8, + 9, + 7	ID card C + tables	**22:1/2** **22:3/4**	Patterns Language	Concept ID card B
4:1/2 **4:3/4**	× 2, × 4 × 10, × 5	× tables × tables	**23:1/2** **23:3/4**	× 6, × 8 × 7	× tables × tables
5:1/2 **5:3/4**	× 10, × 5 × 2, × 5, × 4, × 10, × 0, × 1	× tables × tables	**24:1/2** **24:3/4**	Language × 7, × 8	ID card A × tables
6:1/2 **6:3/4**	+ 8, + 9, + 7 × 4	+ tables × tables	**25:1/2** **25:3/4**	Time Time	Measure Measure
7:1/2 **7:3/4**	Fractions Area	Concept Measure	**26:1/2** **26:3/4**	Language Roman numerals	ID card B Concept
8:1/2 **8:3/4**	Area Near doubling	Measure Concept	**27:1/2** **27:3/4**	Time Roman numerals	Measure Concept
9:1/2 **9:3/4**	Place value Place value	Concept Concept	**28:1/2** **28:3/4**	Division ÷ 2, ÷ 4	Concept ÷ tables
10:1/2 **10:3/4**	Chance Fractions	Chance Concept	**29:1/2** **29:3/4**	Division North, south, east, west	Concept Concept
11:1/2 **11:3/4**	Chance − 5, − 10, − 4	Chance − tables	**30:1/2** **30:3/4**	÷ 5, ÷ 10 × 6, × 7	÷ tables × tables
12:1/2 **12:3/4**	Flip, slide and turn Place value	Concept Concept	**31:1/2** **31:3/4**	× 4, × 6, × 7, × 8, × 9 Division	× tables Concept
13:1/2 **13:3/4**	× 2, × 4, × 8, × 5, × 10 The jump strategy	× tables Strategy time	**32:1/2** **32:3/4**	÷ linked with × Length	Concept Measure
14:1/2 **14:3/4**	Language × 4, × 8	ID card C × tables	**33:1/2** **33:3/4**	× 6, × 9 Language	× tables ID card B
15:1/2 **15:3/4**	Addition linked to subtraction − 8, − 9	Concept − tables	**34:1/2** **34:3/4**	÷ 3, ÷ 6 Rounding off money	÷ tables Concept
16:1/2 **16:3/4**	× 3, × 6 Finding change	× tables Concept	**35:1/2** **35:3/4**	Crossnumber puzzle ÷ 5, ÷ 10	Concept ÷ tables
17:1/2 **17:3/4**	× 3, × 6 Chance	× tables Chance	**36:1/2** **36:3/4**	Problem solving Rounding off money	Strategy time Concept
18:1/2 **18:3/4**	× 4, × 8 × 3, × 6	× tables × tables	**37:1/2** **37:3/4**	Language Personal measures	ID card A Measure
19:1/2 **19:3/4**	× 9 × 9	× tables × tables	**Answers**	These can be found in the middle of this book on pages A1 to A15.	

Arithmetic card

	A	B	C	D	E	F	G	H	I	J
1	7	3	11	20	4	2	7	50	37	$6
2	4	7	17	16	25	10	11	80	93	$1
3	10	1	14	12	64	14	3	20	16	$7
4	3	5	19	17	16	4	15	70	55	$3
5	5	8	16	13	81	12	19	40	100	$8
6	8	6	12	18	1	16	1	60	71	$5
7	2	10	18	11	36	8	9	100	48	$9
8	6	4	15	14	9	20	17	30	82	$2
9	1	9	20	19	100	18	5	90	64	$10
10	9	2	13	15	49	6	13	10	29	$4

How to use this card

If students were told to 'subtract B from C', they would write:

❶ 11 – 3 = 8 ❷ 17 – 7 = 10 ❸ 14 – 1 = 13 ❹ 19 – 5 = 14 ❺ 16 – 8 = 8
❻ 12 – 6 = 6 ❼ 18 – 10 = 8 ❽ 15 – 4 = 11 ❾ 20 – 9 = 11 ❿ 13 – 2 = 11

Other instructions might be:

- **Multiply column B by 4.**
- **Add columns A and C.**
- **Halve column F.**
- **What multiplied by 10 gives column H?**
- **What is the change from $10 if I spent what it is in column J?**
- **Double column G.**
- **Subtract column B from column H.**
- **What must be added to column I to make 100?**
- **Multiply column A by 5.**

The applications of this card are endless.

ID card A

Do not write on this card.

1
m
stands for
m__________

2
cm
stands for
c__________

3
mm
stands for
m__________

4
m^2
stands for s_______
m__________

5
cm^2
stands for s_______
c__________

6
L
stands for
l__________

7
mL
stands for s
m__________

8
kg
stands for
k__________

9
g
stands for
g__________

10
h or hr
stands for
h__________

11
min
stands for
m__________

12
s
stands for
s__________

13
°C
means
d______ C______

14
0 °C
is the f__________
p______ of w______

15
100 °C
is the b__________
p______ of w______

16
am
means
b______ n______

17
pm
means
a______ n______

18

d______ watch

19

a__________
clock

20
5 groups of 7
means
5 ______ 7

21
Share 20 among 5.
means
20 ______ 5

22
How many groups of 6 in 18?
means
18 ______ 6

23
1, 2, 3, …
are the
c__________
numbers

24
2, 4, 6, …
are the
e__________
numbers

25
1, 3, 5, …
are the
o__________
numbers

26
1st, 2nd, …
are the
o__________
numbers

27
493 872
has 6
d__________

28
Balls Lost
May
June
July
stands for one ball.
p__________ graph

29
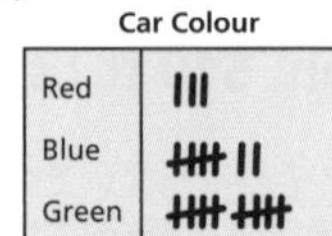

This is a t__________.

30
Colour of Dress
5
4
3
2
1
A B C
c__________ graph

See page A1 for answers.

 • *AUSTRALIAN SIGNPOST MATHS 4 MENTALS* • ISBN 978 0 6557 0884 1

ID card B

Do not write on this card.

1 h____________ line

2 v____________ line

3 p____________ lines

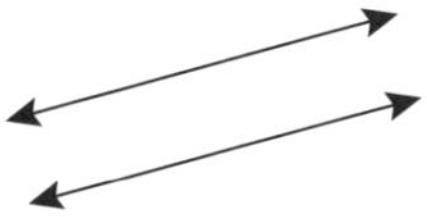

4 p____________ lines

5 0 1 2 3 4 n____________ line

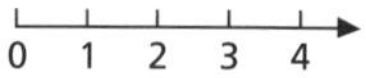

6 a______ of s____________

7 t____________

8 f________ or r__________

9 slide or t__________

10 t________ or r__________

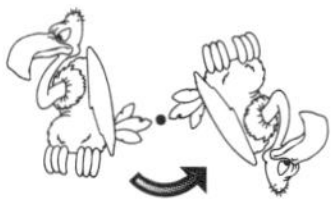

11 a________ angle

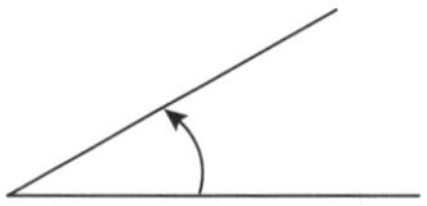

12 r________ angle

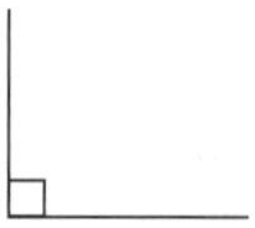

13 s__________ angle

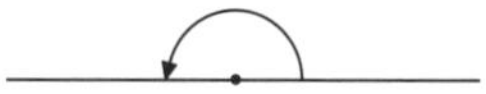

14 o________ angle

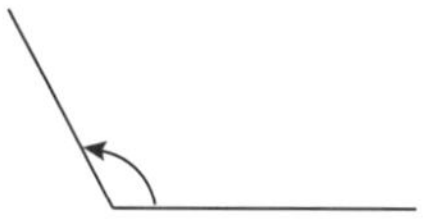

15 r__________ angle

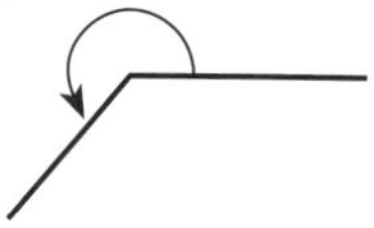

16 r____________

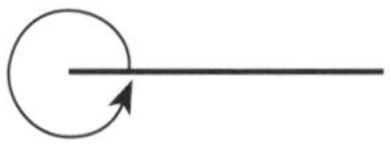

17 v________ of an angle

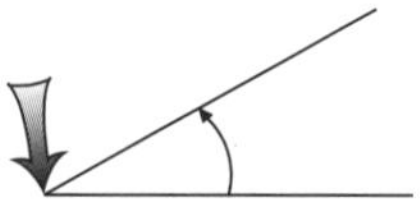

18 a______ of an angle

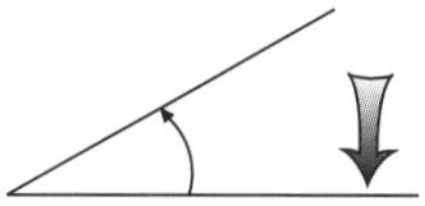

19 7 2 5 tens 9 ones is a n__________ expander

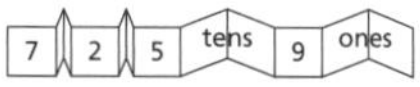

20 H T U a __________

21 _____ sixths shaded

22 ___ hundredths shaded

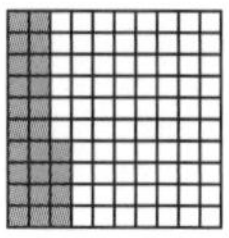

23 ___ tenths shaded

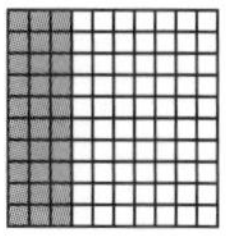

24 0· ___ shaded

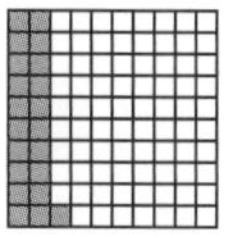

25 ___ hundredths covered

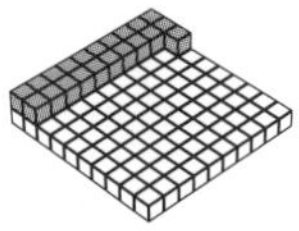

26 0·5 decimal p__________

27

February						
Sun	Mon	Tue	Wed	Thu	Fri	Sat
1	2	3	4	5	6	7
8	9	10	11	12	13	14
15	16	17	18	19	20	21
22	23	24	25	26	27	28

c __________

28 c __________

29 4 3 2 1 0 A B C D E F The c__________ of the dot are E2

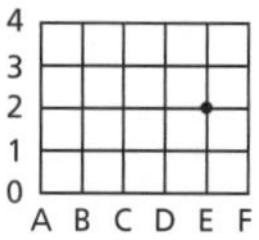

30 1m 1m 1 square m__________

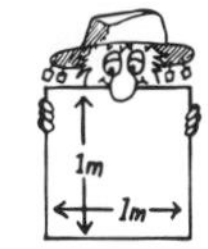

See page A1 for answers.

ID card C

Do not write on this card.

1 o______	2 t______	3 s______	4 r______	5 r______
6 t______	7 p______	8 q______	9 p______	10 h______
11 o______	12 k______	13 r______ shapes	14 i______ shapes	15 d______
16 f______	17 c______ or v______	18 e______	19 b______	20 f______ surface
21 c______ surface	22 s______	23 c______	24 c______	25 c______
26 p______	27 p______	28 n______ of a cube	29 net of a s______ p______	30 net of a t______ p______

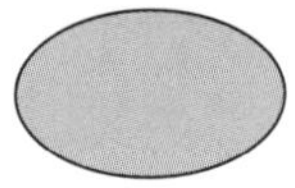

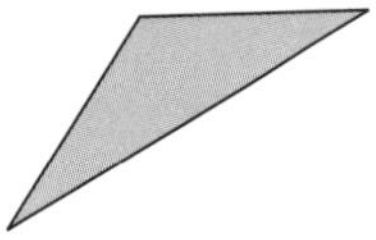

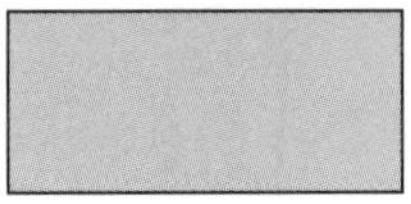

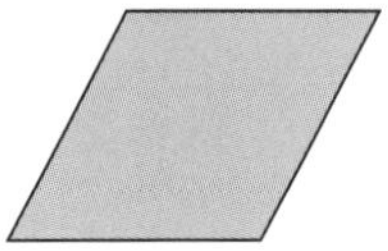

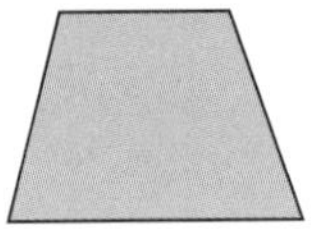

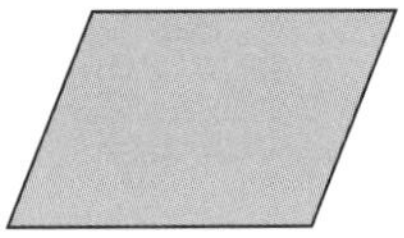

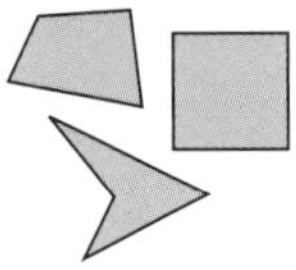

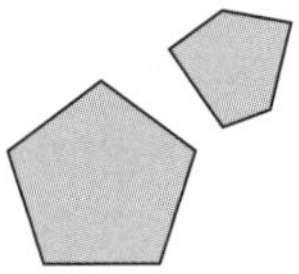

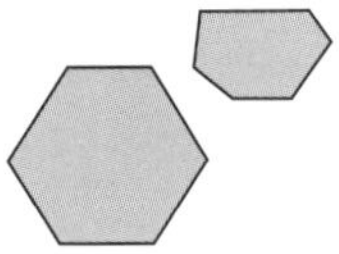

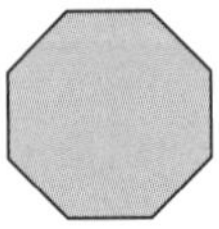

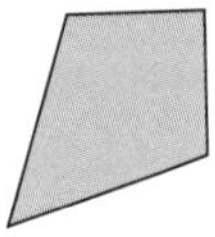

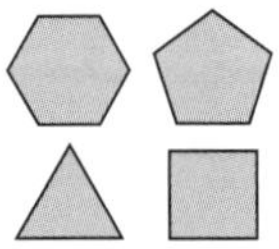

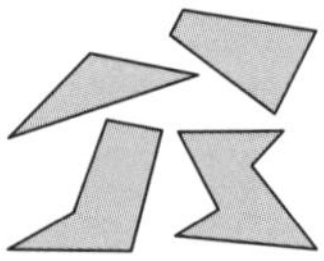

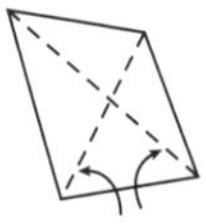

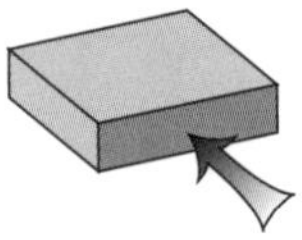

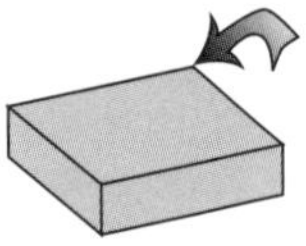

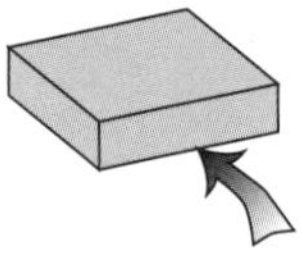

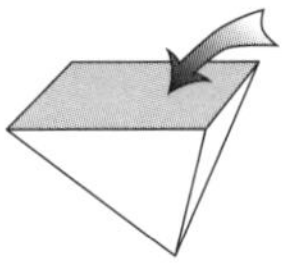

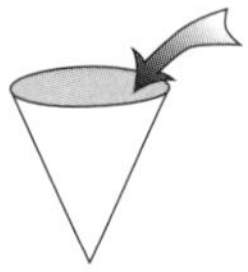

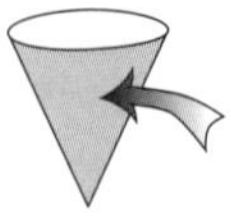

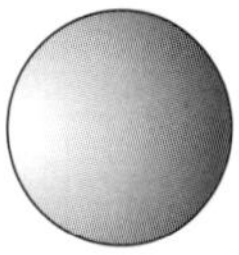

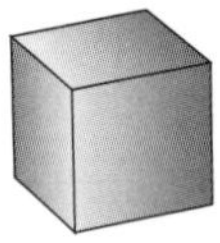

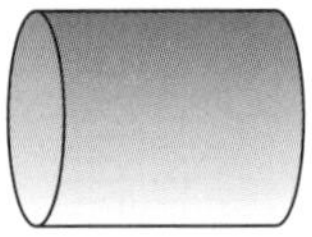

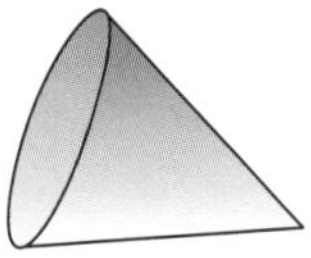

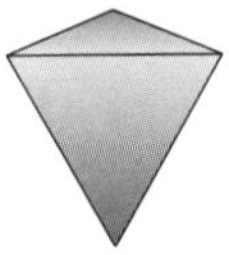

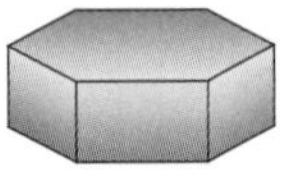

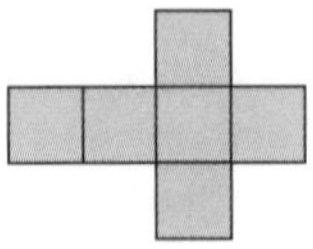

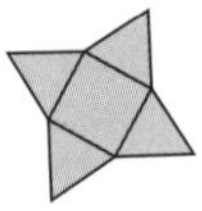

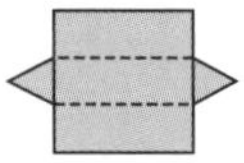

See page A1 for answers.

Examples of measurements

1

2

3

4

5

6

1
- The **width of the boy's finger** is about **1 cm**.
- The **length of a place-value tens block** is **10 cm**.

2
- The **height of the girl** is a little more than **1 m**.

3
- The **container of milk** holds **2 L**.
- The **can of Fizz** holds **375 mL**.
- The **teaspoon** holds **5 mL**.

4
- The **boy** has a mass of **40 kg**.
- The **margarine** has a mass of **500 g**.

5
- The area of the **window** is about **2 m²**.
- The **top of a place-value ones block** is **1 cm²**.

6
- **30°C** is a **hot** day.
- **3°C** is a very **cold** day.

Use the pictures above to estimate the answers to these questions.

1 a How high is the glass?
b How wide is the table?

2 a How wide is the clothes line?
b How tall is the woman?

3 a How much will the bucket hold?
b How much will the cup hold?

4 a What is the mass of the dog?
b What is the mass of 2 L of milk?

5 a What is the area of the table?
b What is the area of the top of a matchbox?

6 a What is the temperature on a very hot day?
b What is the temperature on a cool day?

AUSTRALIAN SIGNPOST MATHS 4 MENTALS • ISBN 978 0 6557 0884 1

1:1 ☐ out of 17

1. 2 + 3 ______
2. 7 × 2 ______
3. 57 − 10 ______
4. 3 × 5 ______
5. 54 + 35 = ______
6. 12 less than 34 ______
7. Half of 22 ______
8. 35 + 10 ______
9. 43 + 7 ______
10. 23 + 61 = ______
11. 40 000 + 2000 + 900 + 20 + 6 ______
12. $\frac{1}{10}$, $\frac{2}{10}$, $\frac{3}{10}$, ☐, ☐, ☐, ☐, ☐
13. Circle the largest number.

    46 716,   46 293,   44 912
14. If 36 + 7 = 43, then 136 + 7 = ______
15. Colour the change I would get from 50 cents when I spend 35 cents.

16. Which of the terms *impossible*, *unlikely*, *likely* or *certain* describes the chance of you finding $20 tomorrow? ______
17. Match each fraction to a decimal.

$\frac{7}{10}$	0·9
$\frac{4}{10}$	0·7
$\frac{9}{10}$	0·4

$1\frac{8}{10}$	1·6
$1\frac{6}{10}$	1·8
$1\frac{3}{10}$	1·3

1:2 ☐ out of 15

1. 20 + 13 ______
2. 5 × 4 ______
3. 36 − 19 ______
4. 58 − 19 ______
5. 46 + 33 = ______
6. 13 + ______ = 20
7. 5 + ______ = 20
8. 647 − 8 ______
9. 522 − 6 ______
10. 34 + 42 = ______
11. Write the decimal for 7 tenths. ______
12. Use the jump strategy to find:

    36 + 29 = ______

13. Use decimals to write zero point nine. ______
14. What date is the:

    a second Saturday? ______

    b fourth Monday? ______

DECEMBER

Sun	Mon	Tue	Wed	Thu	Fri	Sat
		1	2	3	4	5
6	7	8	9	10	11	12
13	14	15	16	17	18	19
20	21	22	23	24	25	26
27	28	29	30	31		

15. Write these numbers on the place-value chart.

    a 846 393   **b** 80 475   **c** 87 637

	Thousands	Hundreds	Tens	Ones
a				
b				
c				

\+ tables

+ 2	+ 3	+ 4
0, 4, 7, 2, 6, 10, 3, 9, 5, 8	8, 0, 3, 9, 2, 7, 10, 4, 6, 5	8, 0, 3, 9, 2, 7, 10, 4, 6, 5

1:3

out of 8

1 Write the decimal for:

a $\frac{8}{10}$ ______ b $\frac{2}{10}$ ______ c $\frac{7}{10}$ ______

d $8\frac{6}{10}$ ______ e $3\frac{5}{10}$ ______

2 Colour 3 fifths of this shape.

3 The fraction for 5·8 is $5\frac{8}{10}$.

Write the fraction for:

a 1·8 $\frac{\square}{\square}$ b 3·9 $\frac{\square}{\square}$ c 6·7 $\frac{\square}{\square}$

4 Colour the change you would get from $2 when you spend $1.35.

5 Sarah drew 4 monsters.
She gave each 5 legs.
How many legs were there altogether? ______

6 Bridge to the next ten to find:

a 57 + 8 ______

b 36 + 7 ______

c 69 + 9 ______

d 35 + 8 ______

7 This is the net of a ______.

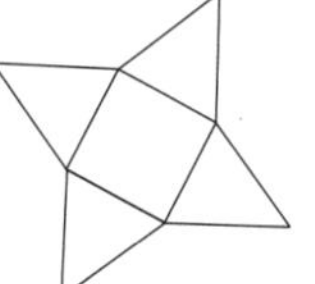

8 How many digits in 846 901? ______

1:4

Extension

out of 6

1 a How many 50c coins make $2? ______

b How many 20c coins make $2? ______

c How many 10c coins make $2? ______

d How many 5c coins make $2? ______

2 Is $8\frac{3}{5}$ equal to $8\frac{4}{10}$? ______

3 Jane and I collected cards.
She collected 35 more than me.
If I collected 145 cards, how many cards do we have altogether? ______

4 Jonkey hit 35 golf balls and Scott hit 6 less. How many did they hit altogether? ______

5

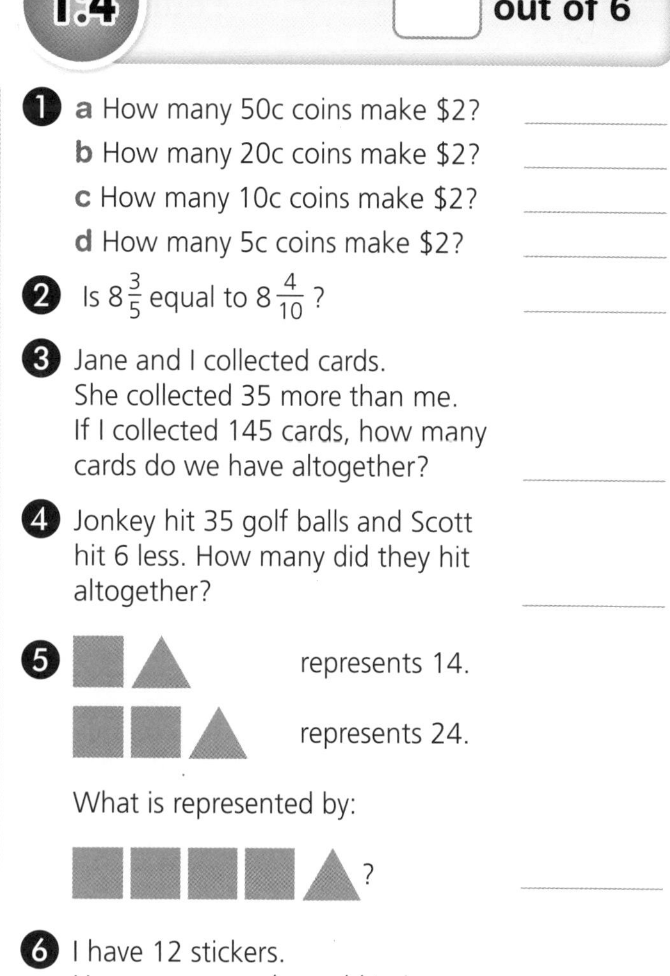

■▲ represents 14.

■■▲ represents 24.

What is represented by:

■■■■▲? ______

6 I have 12 stickers.
How many people could I give:

a 4 stickers? ______ b 6 stickers? ______

Challenge

Write what you know about the number 426 930.

______

______

______

______

______

Measure

Fill out this table about yourself, a relative or a friend.

Name: ______ **Date:** ______

Age: ______	Mass: ______ kg	Shoe size: ______
Height: ______ cm	Waist: ______ cm	Neck size: ______ cm

  • ISBN 978 0 6557 0884 1

2:1 out of 18

1. 20 + 25 ______
2. 50 − 30 ______
3. 40 + 37 ______
4. 60 − 26 ______
5.
$$\begin{array}{r} 53 \\ +32 \\ \hline \end{array}$$
6. 8 less than 100 ______
7. 8 groups of 10 ______
8. 14 + ______ = 20
9. 12 + ______ = 30
10.
$$\begin{array}{r} 75 \\ -42 \\ \hline \end{array}$$
11. Name each shape and write the number of faces.

a 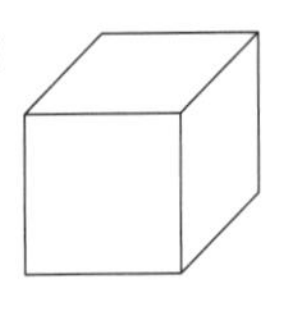

______

______

b 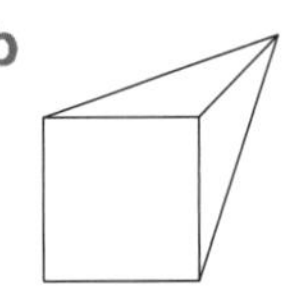

______

______

12. Write these decimals as tenths.

a 0·3 $\frac{}{10}$ b 0·6 $\frac{}{}$ c 0·9 $\frac{}{}$

13. I get $5 pocket money each week. How much will I get in 4 weeks? ______
14. a 53, 63, 73, ______, ______, ______, ______, ______
    b 35, 40, 45, ______, ______, ______, ______, ______
    c 530, 540, 550, ______, ______, ______, ______
    d 28, 26, 24, ______, ______, ______, ______, ______
15. Is 32 520 larger than 32 509? ______
16. 30 000 + 2000 + 200 + 40 + 2 ______
17. I walked along a 5 m balance beam 3 times. How far did I walk? ______
18. The number before 56 723 is ______.

2:2 out of 17

1. 42 + 36 ______
2. 83 − 21 ______
3. 28 + 51 ______
4. 70 − 22 ______
5.
$$\begin{array}{r} 34 \\ +34 \\ \hline \end{array}$$
6. 19 subtract 12 ______
7. 27 minus 15 ______
8. 12 shared by 3 ______
9. 70 divided by 10 ______
10.
$$\begin{array}{r} 64 \\ +23 \\ \hline \end{array}$$
11. 

    a This time is ______ minutes past ______.
    b It is also read as ______.

12. a 

    b

    a: 2 : ______ ; ______ past 2

    b: 7 : ______ ; ______ to 8

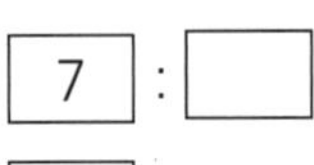

13. This is a ______. It has ______ faces, ______ edges and ______ corners. The cross-section is a ______.

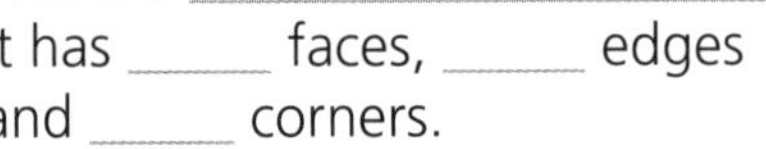

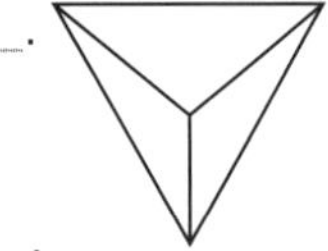

14. Is the height of your mother more than 2 metres? ______
15. 8 days after Wednesday is ______.
16. The 9th month of the year is ______.
17. 13 + 17 + 12 + 8 ______

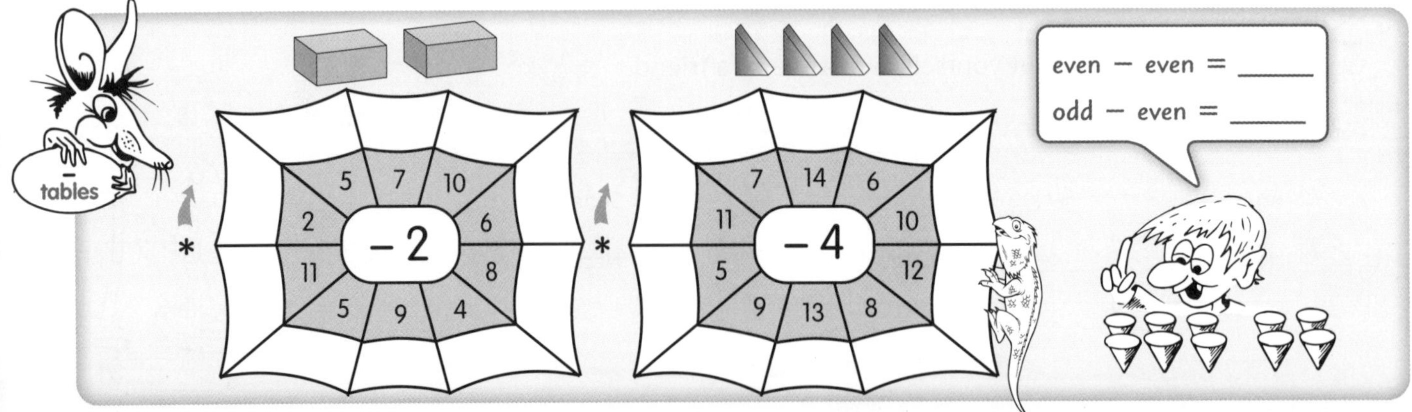

2:3 out of 10

1. Round 4583 to the nearest 1000. ______
2. I had 34 balls and lost some.
   How many did I lose if I have 18 left? ______
3. Write the numeral for sixty-seven thousand, 4 hundred and seventeen. ______
4. Describe this rectangle and write the area.

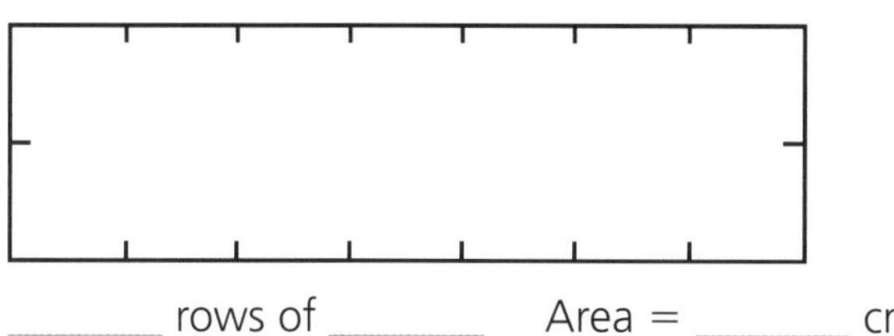

______ rows of ______ Area = ______ cm^2

5. **a** The analog time is ______ to ______.

   b The digital time is ______ : ______.

6. What is the time ten minutes after:
   a quarter to 4? ______
   b 3 minutes to 7? ______
7. Give a rule for this pattern.
   20, 40, 60, 80, 100,... ______
8. The number before 6493 ______
9. Cross out the mistake in the pattern.

10. Write in short form.
    a 60 grams ______ **b** 64 kilograms ______

2:4 Extension out of 9

1. 100 − 15 − 15 − 15 − 15 − 15 ______
2. I own 43 stickers. This is 8 more than Rhonda owns.
   How many do we own altogether? ______
3. In a volleyball game, each team has 6 players on the court. How many players would be in 10 games altogether? ______

4. 25 + 35 + 45 + 55 ______
5. 90 − 25 − 25 − 25 ______
6. What is the time 35 minutes after:
   a 5:35? ______ **b** 8:52? ______
7. How many weeks in 3 years? ______
8. 207 more than 198 is ______.
9. The date is May 14 and my birthday is May 30. What day will my birthday be if it is Sunday today? ______

Challenge

Write questions that are equal to:

a 34 − 12	**b** 35 + 3	**c** 5 × 8
= ______	= ______	= ______
= ______	= ______	= ______
= ______	= ______	= ______
= ______	= ______	= ______
= ______	= ______	= ______

+ tables

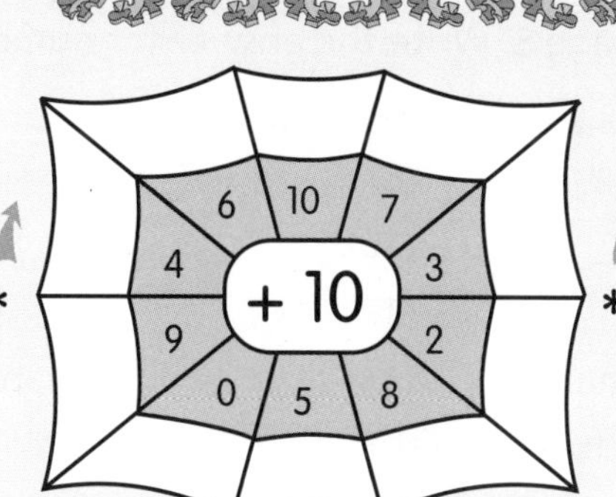

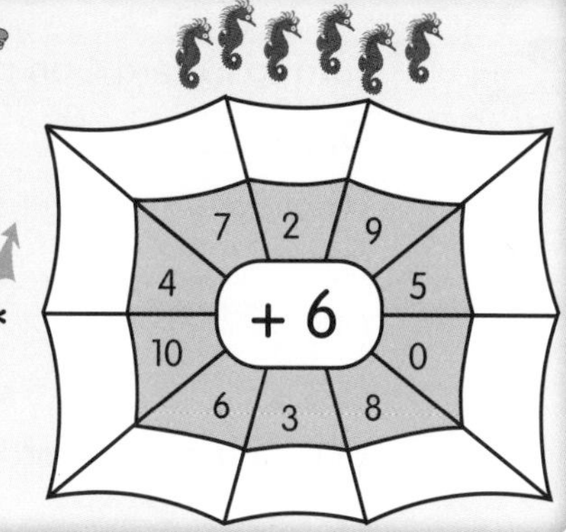

 • *AUSTRALIAN SIGNPOST MATHS 4 MENTALS* • ISBN 978 0 6557 0884 1

3:1 out of 19

1. 35 + 10 ____
2. 2 × 6 ____
3. 4 × 6 ____
4. 49 − 19 ____
5. $\begin{array}{r} 57 \\ +21 \\ \hline \end{array}$
6. Digits in 53 476 ____
7. 35 + 18 − 18 + 6 ____
8. 32 + 35 − 35 + 8 ____
9. Multiply 5 by 4. ____
10. $\begin{array}{r} 18 \\ +30 \\ \hline \end{array}$
11. Shade 3 eighths.

12. How many school days in 10 weeks? ____
13. The difference between 24 and 17 is ____.
14. Draw two irregular shapes with 6 sides.
15. Jon is now 18. How old will he be in 25 years? ____
16. How many lines of symmetry has a square? ____
17. Write the numeral shown below. ____

2	thousands	4	hundreds	1	tens	8	ones

18. Write the numeral fifteen thousand, seven hundred and forty-eight. ____
19. What time is 15 minutes after 5:44? ____

3:2 out of 18

1. 16 + 47 ____
2. 7 × 10 ____
3. 6 × 5 ____
4. 9 × 2 ____
5. $\begin{array}{r} 79 \\ -26 \\ \hline \end{array}$
6. Divide 25 by 5. ____
7. \$4.75 − \$1.15 ____
8. 35 ÷ 7 × 7 ____
9. 1 m − 40 cm ____
10. $\begin{array}{r} 81 \\ -48 \\ \hline \end{array}$
11. Write the numeral shown below. ____

9		3	hundreds	7	tens	4	ones

12. Write the short form for 400 millilitres. ____
13. 100 more than 56 846. ____
14. Shade $\frac{7}{10}$.

15. Write the number that has 415 hundreds and 34 ones. ____
16. The value of the 7 in 57 203 is ____.
17. Which of the objects: A B C D
   a will roll? ____
   b will stack easily? ____
   c has the most edges? ____
   d has the most surfaces? ____
18. Millilitres in 8 L. ____

ID Card C

Turn to ID card C on page 8. Write the answer for number:

(5) ____ (6) ____
(7) ____ (8) ____
(9) ____ (10) ____
(11) ____ (12) ____
(13) ____ shapes (14) ____ shapes

Make up a study card for any mistakes.

Make a study card: Put questions on one side and answers on the other.

6 [trapezium shape] 9 [pentagon shape] | 6 trapezium 9 pentagon

 • *AUSTRALIAN SIGNPOST MATHS 4 MENTALS* • ISBN 978 0 6557 0884 1

3:3 ☐ out of 8

1 Show each time on the clock.

a 8 to 4 b 13 to 11

2 Write the short form for:

a 9 litres ______ b 23 kilograms ______

c 13 metres ______ d 14 millimetres ______

e 13 square centimetres ______

3 How many 2·5 cm lengths could you cut from a ribbon this long? ______

4 Circle 6 out of 12.

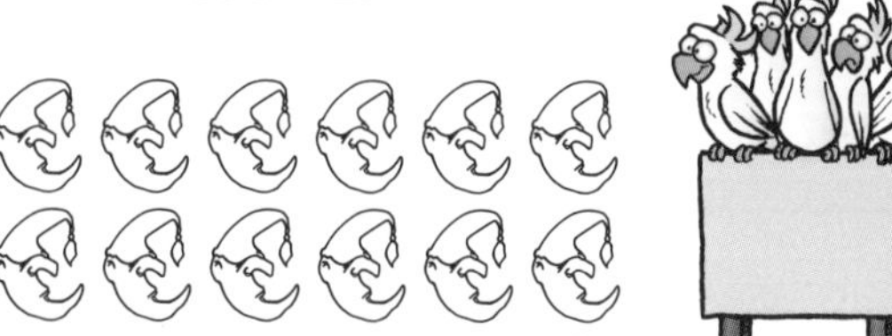

Write the fraction. ______

5 Write the numeral shown below. ______

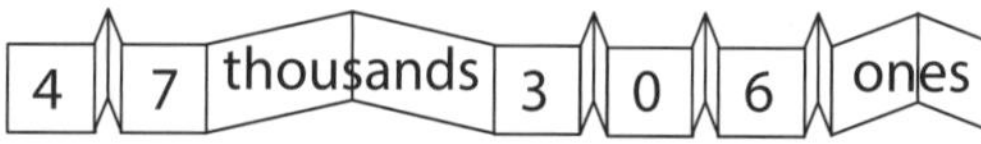

6 Write the number that has:

a 6378 hundreds and 18 ones. ______

b 6306 hundreds and 2 ones. ______

7 How many 10c coins in $3? ______

8 From $10.00 take away $5.60. ______

3:4 Extension ☐ out of 4

1 I had 45 apples. I ate some and had 28 left. How many have I eaten? ______

2 a Colour one half red.

b Colour one quarter blue.

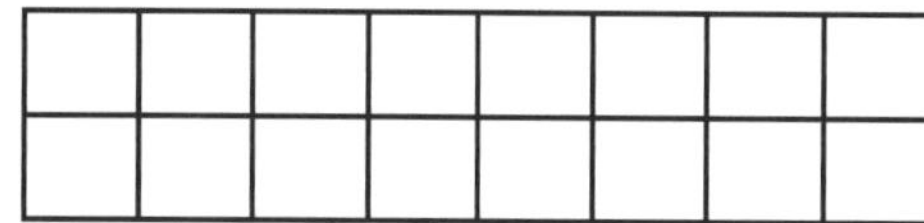

c What fraction is not coloured? $\frac{\square}{\square}$

3 If this pattern continues, how many squares would be in:

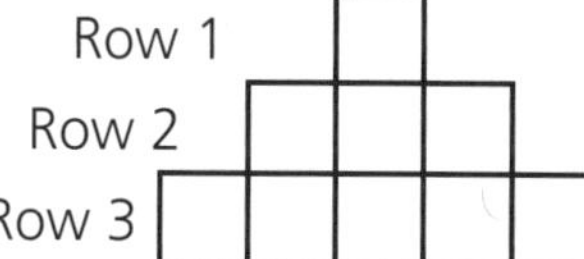

a Row 5? ______ b Row 10? ______

4

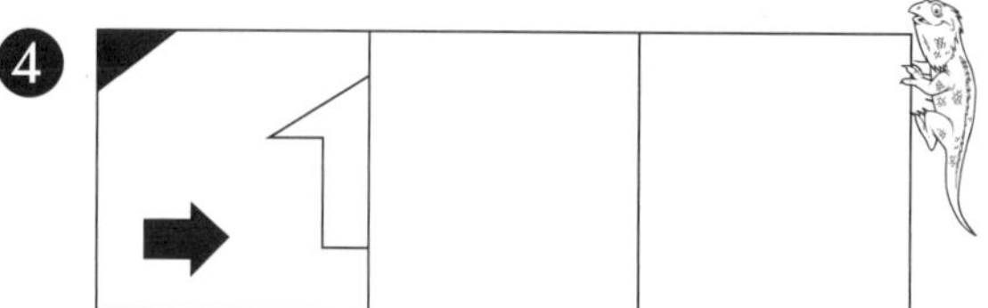

Complete this pattern by flipping each tile to the right.

Challenge

Draw and label fractions of your own.

+ 7	9	3	6	0	4	8	10	2	7	5

 • *AUSTRALIAN SIGNPOST MATHS 4 MENTALS* • ISBN 978 0 6557 0884 1

4:1 out of 18

1. 29 + 30 ______
2. 76 − 40 ______
3. 6 × 10 ______
4. 8 × 10 ______
5. $\begin{array}{r} 87 \\ -36 \\ \hline \end{array}$
6. 26 + ______ = 30
7. 58 + ______ = 60
8. Digits in 54 786 ______
9. 8 × 5 ______
10. $\begin{array}{r} 59 \\ -27 \\ \hline \end{array}$
11. Write the numeral for ninety-eight thousand, four hundred and eleven. ______
12. How many seconds in 1 minute? ______
13. 80 000 + 5000 + 300 + 80 + 2 ______
14. Round 633 to the nearest hundred. ______
15. **a** 3, 6, 9, ______, ______, ______, ______, ______
    b 4, 8, 12, ______, ______, ______, ______, ______
    c 9, 18, 27, ______, ______, ______, ______, ______
16. Put these numbers in order, from smallest to largest: 4265, 4246, 4286.
    ______
17. Three groups of 10c = ______ 
18. 9 bones were shared between 3 dogs. How many would each dog get? ______

4:2 out of 17

1. 26 + 9 ______
2. 5 × 3 ______
3. 5 × 6 ______
4. 2 × 5 ______
5. $\begin{array}{r} 79 \\ -39 \\ \hline \end{array}$
6. Digits in 403 729 ______
7. 9 groups of 2 ______
8. 7 rows of 5 ______
9. 3 times 10 ______
10. $\begin{array}{r} 49 \\ -28 \\ \hline \end{array}$
11. Write the numeral shown. ______

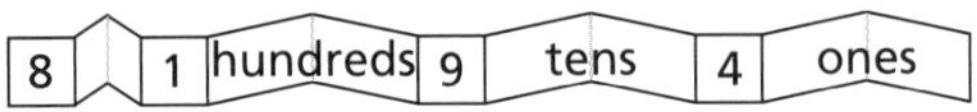

12. **a** 30, 34, 38, ______, ______, ______, ______
    b 18, 20, 22, ______, ______, ______, ______
13. Circle numbers that round to 60 000.

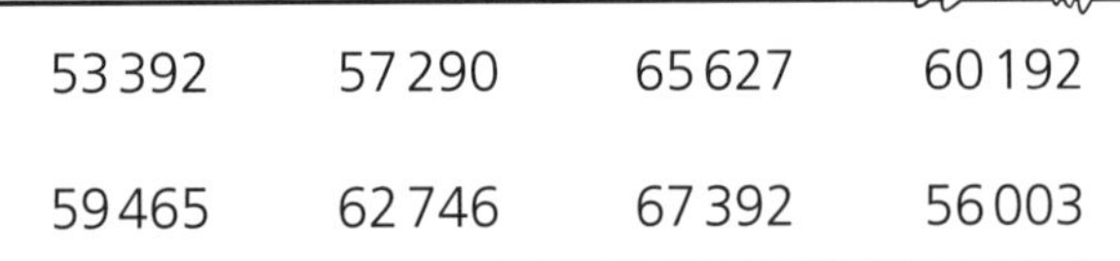

53 392	57 290	65 627	60 192
59 465	62 746	67 392	56 003

14. Three groups of 5c = ______ 
15. These are shared by four people.

One share = ______

16. Round 456 to the nearest hundred. ______
17. 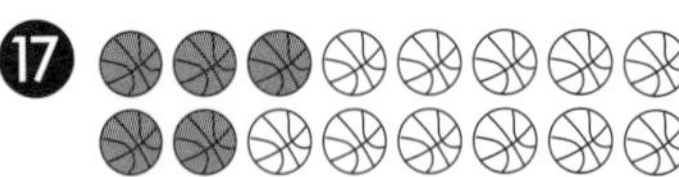

What fraction of this group is:
a shaded? ______ **b** unshaded? ______

× tables

3	5	7	10	8	2	6	4	9	0

* × 2

* × 4

Double × 2 answers to find × 4 answers.

 ISBN 978 0 6557 0884 1

4:3 ☐ out of 11

1. Write the numeral shown. ________

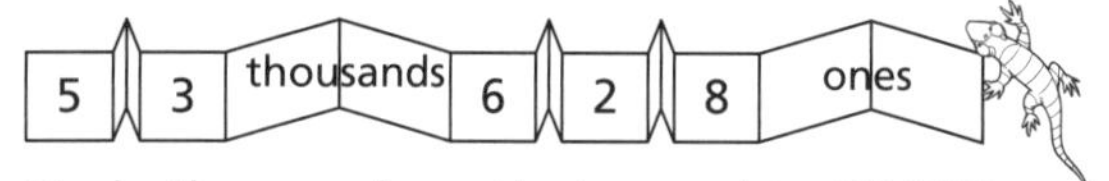

2. Circle the numbers that round to 80 000.

76 483	86 372	79 802	75 000
89 374	71 356	76 394	84 476

3. a 6, 12, 18, ____, ____, ____, ____, ____

   b 7, 14, 21, ____, ____, ____, ____, ____

   c 35, 40, 45, ____, ____, ____, ____

4. Months in 2 years. ________

5. 2 groups of 2 = ________

   3 groups of 2 = ________

6. How many digits in 5304? ________

7. Add 45 hundreds, 6 tens and 7 ones. ________

8. Write 5309 in words. ________

9. Round 8460 to the nearest hundred. ________

10. What is 10 more than 3578. ________

11. Use the numeral expanders to write 5745.

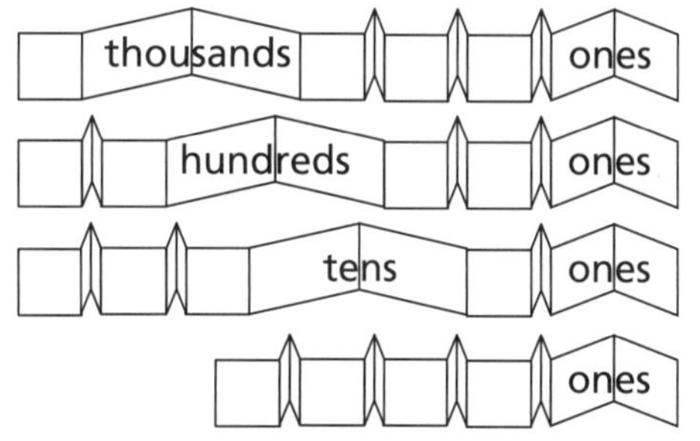

4:4 ☐ out of 10

Extension

1. a 13, 26, 39, ____, ____, ____, ____

   b 15, 30, 45, ____, ____, ____, ____

2. a Four groups of 70c = ________

   b Five groups of 70c = ________

3. 17 + 13 + 18 + 12 + 8 = ________

4. a 2 hours before 11:15. ________

   b 2 hours after 11:15. ________

5. $\frac{1}{2} + \frac{1}{2} + \frac{1}{2} + \frac{1}{2} + \frac{1}{2}$ ________

6. I had 18 mandarins. I ate some and had 11 left. How many did I eat? ________

7. 6, 17, 28, ____, ____, ____, ____, ____

8. 90 − 15 − 15 − 15 ________

9. Mary has $20, Jack has $32 and Jill has $23. The total of their money is ________.

   How much more has Jack than Mary? ________

10. How many odd numbers between 8 and 28? ________

Challenge

Write your own number pattern for each rule.

Add 12. ________

Subtract 10. ________

Add 7. ________

Subtract 14. ________

Add 21. ________

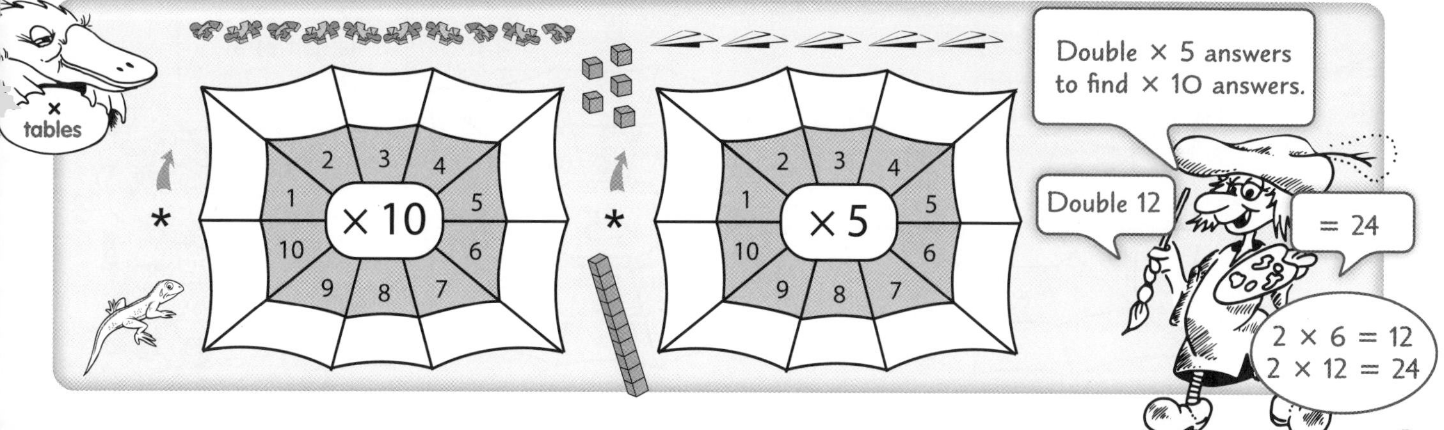

5:1 ☐ out of 17

1. 54 + 6 ______
2. 35 + 5 ______
3. 87 − 3 ______
4. 68 − 5 ______
5. $\begin{array}{r} 83 \\ +15 \\ \hline \end{array}$
6. 5, 10, 15, ______, ______
7. 3, 6, 9, ______, ______
8. 4, 8, 12, ______, ______
9. 23, 33, 43, ______, ______
10. $\begin{array}{r} 62 \\ +35 \\ \hline \end{array}$

11. Round 45 634 to the nearest ten-thousand. ______
12. What fraction of the rabbits are shaded? $\frac{\square}{\square}$

13. a 10, 20, 30, ______, ______, ______, ______
    b 2, 4, 6, ______, ______, ______, ______, ______
14. 3 groups of $2 = $______

15. How many minutes past 4 o'clock?

a  ______

b  ______

16. I shared 12 toys into 3 boxes. How many in each box? ______
17. Double 4 then double your answer. ______

5:2 ☐ out of 16

1. 8 × 5 ______
2. 8 × 2 ______
3. 7 × 10 ______
4. 5 × 5 ______
5. $\begin{array}{r} 83 \\ -27 \\ \hline \end{array}$
6. 6 groups of 5 ______
7. 7 rows of 10 ______
8. Days in 5 weeks ______
9. Months in 2 years ______
10. $\begin{array}{r} 41 \\ +54 \\ \hline \end{array}$

11. Circle numbers that round to 50 000.

45 830	48 273	42 469	57 283
55 463	51 394	44 657	53 674

12. What fraction of the hippos are:

a black or grey? $\frac{\square}{\square}$ b not black? $\frac{\square}{\square}$

13. a 81, 90, 99, ______, ______, ______, ______
    b 54, 60, 66, ______, ______, ______, ______
14. Colour: a $\frac{4}{6}$ of this shape. b $\frac{2}{6}$ of this shape.

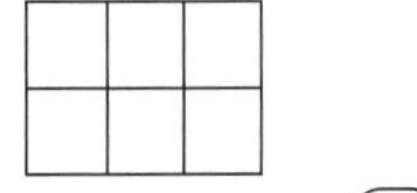

c Write the larger fraction. $\frac{\square}{\square}$

15. What is the time shown on this clock face?

______ past ______

16. Double 6 then double your answer. ______

× tables

×10: 3, 5, 7, 10, 8, 2, 6, 4, 9, 0

×5: 3, 5, 7, 10, 8, 2, 6, 4, 9, 0

3 groups of 5 is half of 3 groups of 10.

 ISBN 978 0 6557 0884 1

5:3 ☐ out of 8

1. 2 groups of 5 = ______
   3 groups of 5 = ______

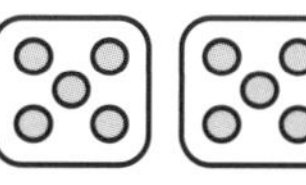

2. Circle numbers that round to 70 000.

71 254	75 048	76 463	64 999
65 488	75 463	72 465	70 896

3. What fraction of this group is:

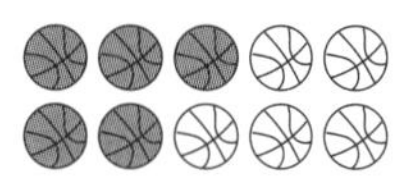

   a shaded? $\frac{\square}{\square}$ **b** not shaded? $\frac{\square}{\square}$

4. The time five minutes before 5:16 is ______.

5. **a** The time is ______.
   b This can also be written as ______.

6. **a** Show the time 25 past 4 on the clock face.
   b This can also be written as ______.

7. **a** How many minutes in 1 hour? ______
   b How many hours in 1 day? ______
   c How many days in 1 week? ______

8. Colour 7 eighths of this circle blue.

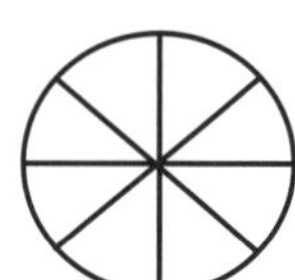

5:4 ☐ out of 8

Extension

1. Shade $\frac{3}{4}$ of this rectangle.

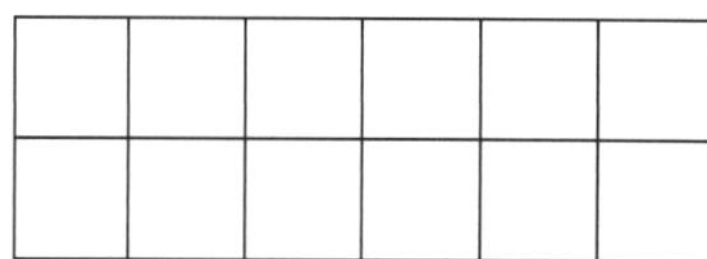

2. **a** What is the time on this clock? ______
   b What is the time 5 minutes before this time? ______

3. Each prawn costs $1.15. How much do 5 prawns cost? ______

4. 4 + 16 + 23 + 7 + 5 ______

5. There are 41 houses in my street. How many have odd numbers? ______

6. 80 stamps shared by 4 people. ______ each

7. Write 6 thousand and nineteen. ______

8. 8 + 8 + 8 + 8 + 8 ______

Challenge

Draw and label large fractions that are less than 1.

a

	2	4	6	8	10
× 2					

b

	2	4	6	8	10
× 5					

c

	1	2	3	4	5
× 4					

d

	1	2	3	4	5
× 10					

e

	6	3	8	7	9
× 0					

f

	4	7	3	9	6
× 1					

6:1 out of 16

1. 1 × 4 ______
2. 4 × 4 ______
3. 6 × 4 ______
4. 3 × 4 ______
5. 86 − 54
6. Double 2 × 4. ______
7. Double 2 × 7. ______
8. Double 2 × 6. ______
9. 2 groups of 4 ______
10. 67 − 23

11. What fraction of the dogs are grey? $\frac{\square}{\square}$

12. Draw the analog time for 7:15 on this clock face.

13. Write as millimetres:

    a 4 cm 3 mm = ______ mm

    b 7 cm 9 mm = ______ mm

14. My box of chocolates held 3 rows of 5 chocolates. How many chocolates does it hold? ______

15. a ______ × 5 = 15 b ______ × 5 = 45

    c ______ × 5 = 50 d ______ × 5 = 35

16.

	7	10	2	5	1	3	8	6	9
× 5									

6:2 out of 17

1. 7 × 4 ______
2. 10 × 4 ______
3. 5 × 4 ______
4. 8 × 4 ______
5. 48 − 27
6. 9 rows of 4 ______
7. Double 2 × 8. ______
8. Double 2 × 10. ______
9. Double 2 × 3. ______
10. 43 − 21

11. 2 groups of 4 = ______

    3 groups of 4 = ______

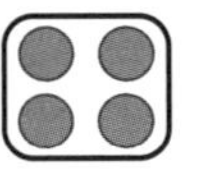

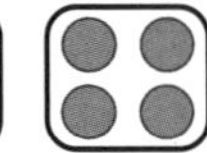

12. What fraction of this shape is shaded? $\frac{\square}{\square}$

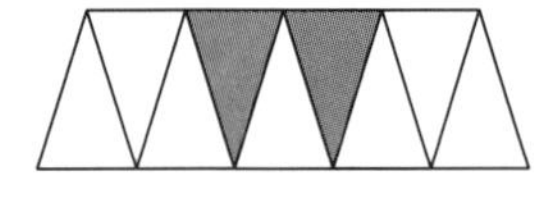

13. The time five minutes before 2:43 is ______.

14. Write the digital time for:

    a quarter past 3

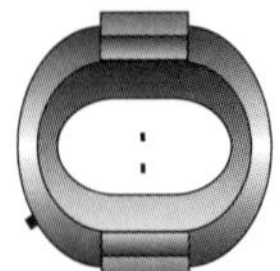

    b half past 3

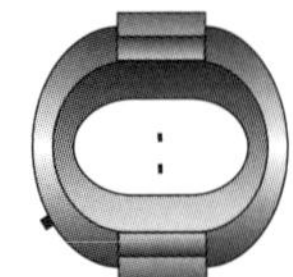

15. This line is ______ mm long.

16. a 12 shared by 3 = ______

    b 12 shared by 4 = ______

17. Draw 3 rows of 4. 3 × 4 = ______

 • *AUSTRALIAN SIGNPOST MATHS 4 MENTALS* • ISBN 978 0 6557 0884 1

6:3 — out of 10

1. Colour 3 tenths of the frogs green and 5 tenths red.

2. Show each time on the clock.

   a 5:53  b 7:28

3. Soccer started at 2:35 and finished at 3:00. For how long did we play? ______

4. This line is ______ centimetres long.

5. Write as centimetres and millimetres.

   a 84 mm = ____ cm ____ mm

   b 59 mm = ____ cm ____ mm

6. 30 000 + 6000 + 200 + 9 = ______

7. a 16 shared by 8 = ____ 16 ÷ 8 = ____

   b 16 shared by 2 = ____ 16 ÷ 2 = ____

8. There were 5 rows of 5 tiles on the wall. How many tiles were there? ______

9.

	7	10	2	5	1	3	8	6	9
× 10									

10. Draw 2 rows of 8 circles. 2 × 8 = ______

6:4 — out of 6 — Extension

1. a 1000 − 101 = ______

   b 1000 − 304 = ______

   c 1000 − 807 = ______

2. How many times would either of the two hands of a clock point to the numeral 3 in one day? ______

3. I worked from 3:12 to 4:00 and then from 4:36 to 5:00. For how long did I work altogether? ______

4. Add 17 each time. 17, ______, ______, ______

5. Colour three quarters of this square.

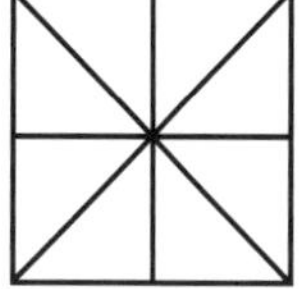

6. a 3 m − 1 cm ______

   b 4 m − 103 cm ______

   c 5 m − 113 cm ______

   d 2 m + 245 cm ______

Challenge

Write your own number patterns and the rule for each.

______

______

______

______

______

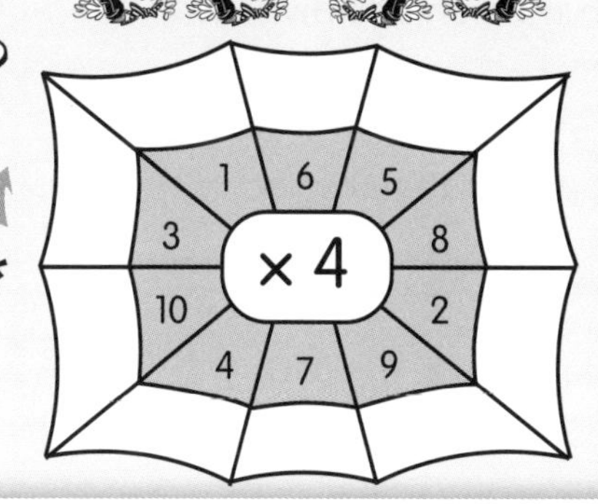

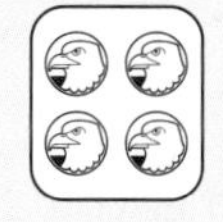

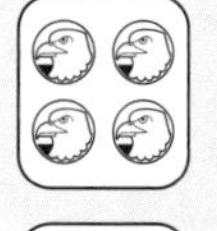

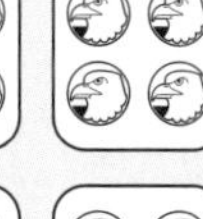

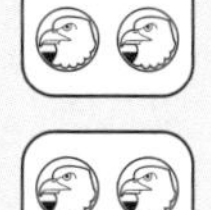

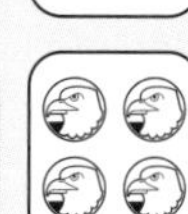

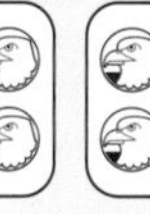

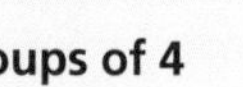

Groups of 4

7:1 ☐ out of 15

1. 6 × 5 ______
2. 7 × 4 ______
3. 9 × 10 ______
4. 3 × 5 ______
5. $\begin{array}{r} 74 \\ -52 \\ \hline \end{array}$
6. 10 more than 573 ______
7. 10 more than 475 ______
8. 4 × 4 + 4 ______
9. 5 × 4 + 4 ______
10. $\begin{array}{r} 89 \\ -56 \\ \hline \end{array}$

11. Write 3 m 25 cm as centimetres. ______
12. Show 3 × 5 below. 3 × 5 = ______

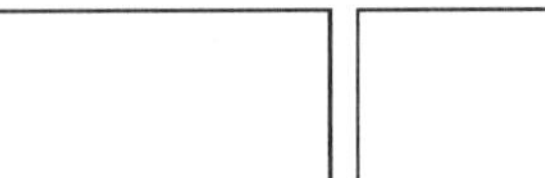

13. **a** How many quarters are coloured? ______

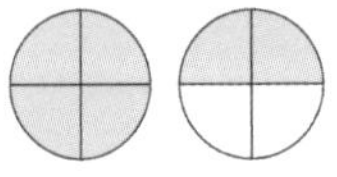

b The mixed numeral for this is .

14. 

Use your ruler to measure:

a the length of this rectangle. ______

b the width of this rectangle. ______

15.   

a This shows ______ groups of ______.

b ______ × ______ = ______

7:2 ☐ out of 16

1. 9 × 4 ______
2. 3 × 4 ______
3. 7 × 4 ______
4. 8 × 4 ______
5. $\begin{array}{r} 42 \\ +46 \\ \hline \end{array}$
6. 5 groups of 10 ______
7. 16 shared by 4 ______
8. 7 rows of 5 ______
9. 10 times 3 ______
10. $\begin{array}{r} 23 \\ +16 \\ \hline \end{array}$
11. Write 2 cm 7 mm as millimetres. ______
12. Estimate then measure this line's length.

Estimate = ______ cm Measure = ______ cm

Write this length in millimetres. = ______ mm

13. Write the improper fraction and mixed number for the part that is shaded.

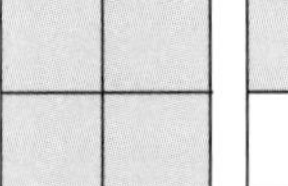 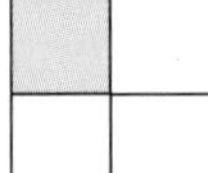  

14. Write as centimetres and millimetres:

a 93 mm ______

b 79 mm ______

15. The area of this shape is ______ cm^2.

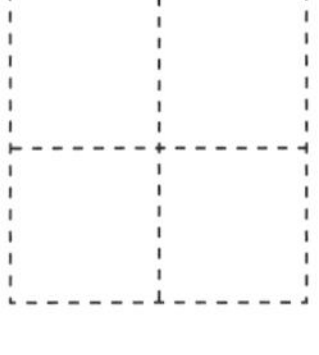

16. Colour five and a half balls.

$\frac{12}{4} = 12 \div 4$

Twelve quarters = 3

Write each fraction as a division.

a $\frac{8}{2}$ = ______ ÷ ______ 

b $\frac{27}{3}$ = ______ ÷ ______ 

In each case, find the answer.

c $\frac{8}{2}$ ______ **d** $\frac{27}{3}$ ______ **e** $\frac{24}{4}$ ______

f $\frac{30}{5}$ ______ **g** $\frac{90}{10}$ ______ **h** $\frac{28}{4}$ ______

7:3 — out of 10

1. Rewrite as metres and centimetres.
   a 459 cm ______
   b 327 cm ______
2. The length of this page correct to the nearest centimetre. ______ cm
3. Write the improper fraction and mixed number for the part shaded.

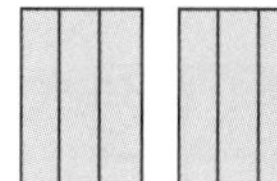
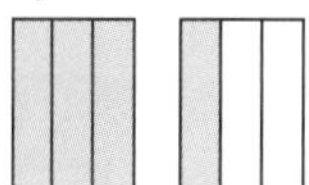

4. Write the mixed number.
   a three and a half ______
   b one and three quarters ______
5. The area of this shape is ______ cm².

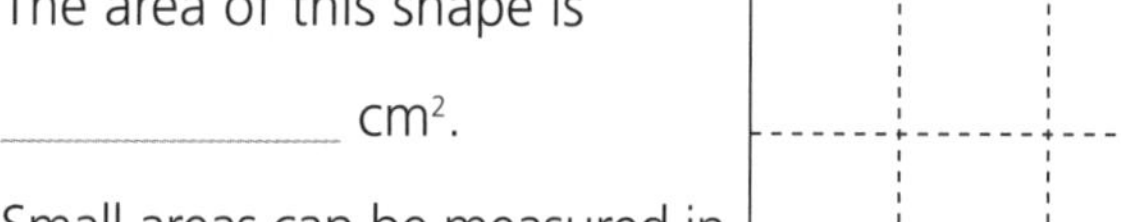

6. Small areas can be measured in s______ c______.
7. Write as millimetres.
   a 9 cm 5 mm ______
   b 4 cm 2 mm ______
8. Write the mixed number for:
   a $\frac{12}{5}$ ______
   b $\frac{8}{3}$ ______
9. 20000 + 200 + 10 + 9 = ______

10.

	7	10	2	5	1	3	8	6	9
× 4									

7:4 — Extension — out of 5

1. a 10 cm take away 1 millimetre. ______
   b 1 cm take away 1 millimetre. ______
2. 9 stamps fit on this square. How many stamps will fit on a square with side lengths twice as long? ______

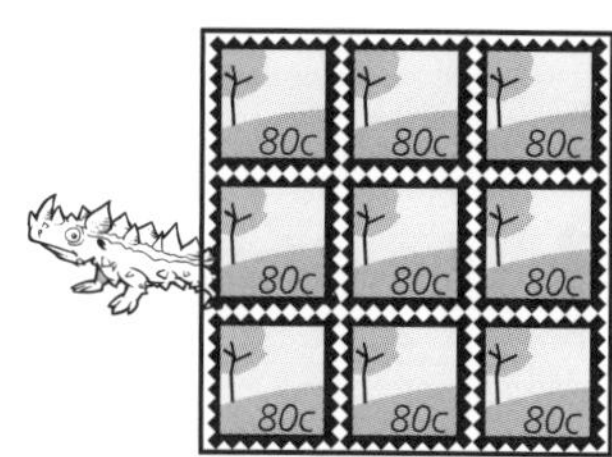

3. What is the time 15 minutes after:
   a 5 to 7? ______ b 6:57? ______
4.

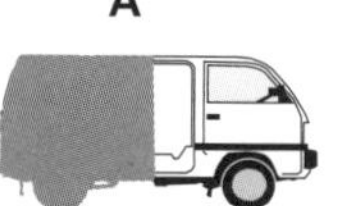

Which van is about $\frac{3}{10}$ covered? ______

5. a 32 + 32 + 32 + 32 ______
   b 32 × 3 ______
   c 32 × 5 ______

Challenge

Draw and label two fractions that are between 1 and 2.

Describe, then find the area of each shape.

a
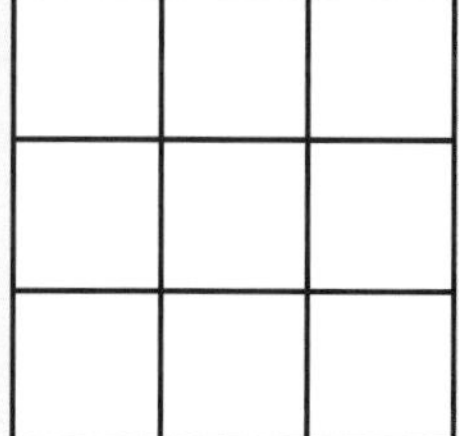

3 rows of ______ = ______ cm²

b
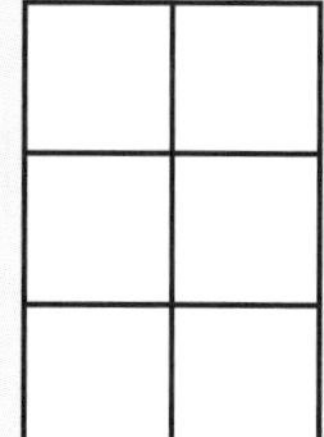

______ rows of ______ = ______ cm²

c
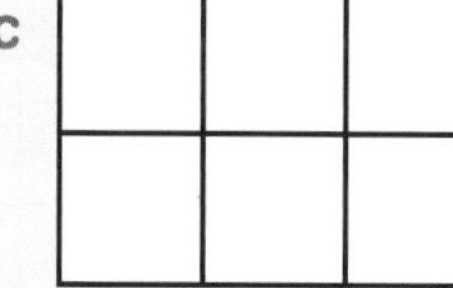

______ rows of ______ = ______ cm²

8:1 ☐ out of 16

1. 23 + 30 ____
2. 27 + 40 ____
3. 49 + 20 ____
4. 56 + 30 ____
5. $\begin{array}{r} 45 \\ +32 \\ \hline \end{array}$
6. 15 + ____ = 20
7. 12 + ____ = 20
8. 40 + 7 ____
9. 30 + 6 ____
10. $\begin{array}{r} 81 \\ +15 \\ \hline \end{array}$
11. Write the improper fraction and mixed number for the part shaded.

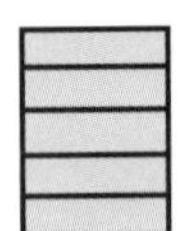 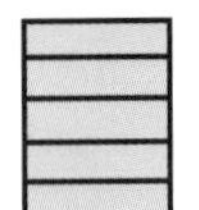 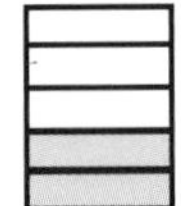  

12. Estimate then measure the length of this line.

Estimate = ____ cm Measure = ____ cm

Write this length in millimetres. = ____ mm

13. $\frac{1}{10}$, $\frac{2}{10}$, $\frac{3}{10}$,      

14. Colour 4 and a half ice creams.

15. a 5 × 4 ____ b 7 × 4 ____

16. The difference between 8 and 19. ____

8:2 ☐ out of 16

1. 46 + 31 ____
2. 35 + 14 ____
3. 36 + 23 ____
4. 34 + 62 ____
5. $\begin{array}{r} 67 \\ -42 \\ \hline \end{array}$
6. 5 + 19 − 19 ____
7. Halve 30. ____
8. 25 less than 56 ____
9. 35 more than 34 ____
10. $\begin{array}{r} 86 \\ -31 \\ \hline \end{array}$

11. How many small triangles cover this area? ____

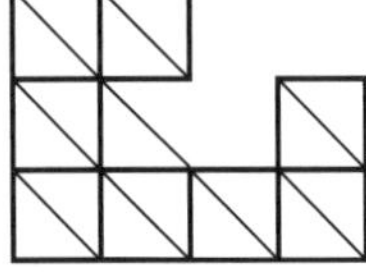

12. Write the improper fraction and mixed number for the part shaded.

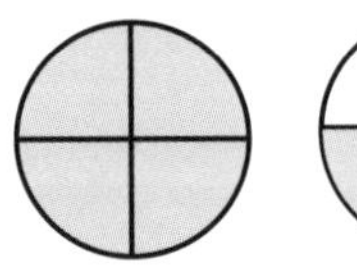 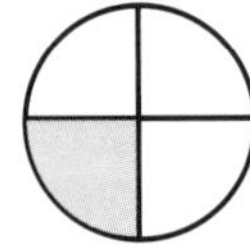  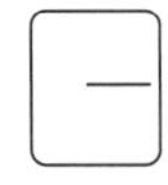

13. The area of this shape is ____ cm^2.

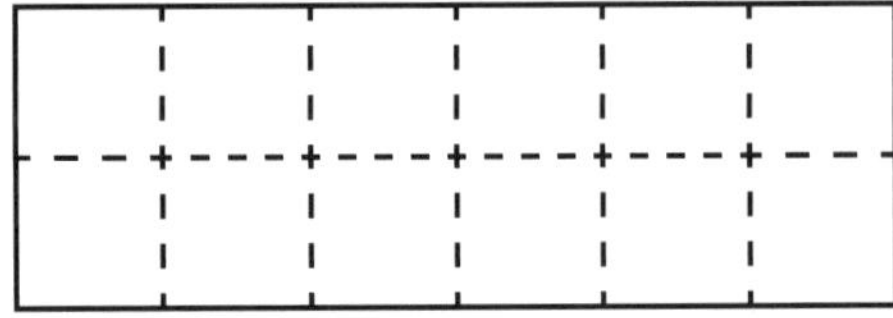

14. Write the mixed number for:
    a $\frac{7}{3}$ ____
    b $\frac{10}{4}$ ____
15. 65 birds. 32 escape. How many are left? ____
16. 34 yellow balls and 54 red balls. How many balls altogether? ____

Find the area of each shape.

a 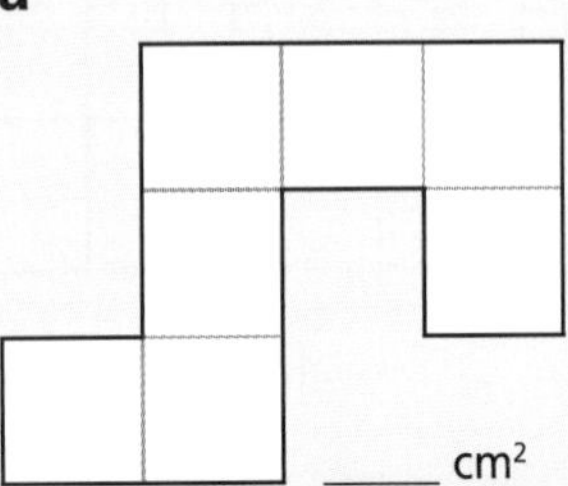 ____ cm^2

b 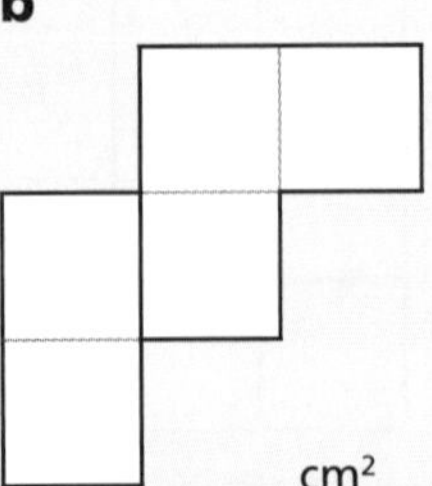 ____ cm^2

c ____ cm^2

 • ISBN 978 0 6557 0884 1

8:3 ☐ out of 7

1.

tens	ones
4	5
+ 3	2

2.

tens	ones
6	2
− 3	1

3. Write the improper fraction and mixed number for the part shaded.

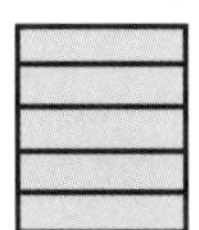
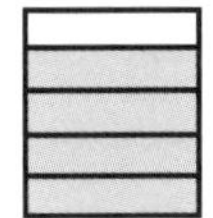
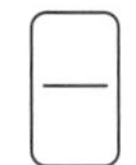
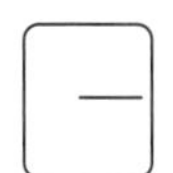

4. True or false? The area of your thumb-nail is about 1 cm^2. ______

5. Which picture is:

a a flip? ______ **b** a slide? ______

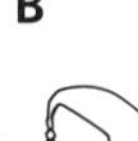

6.

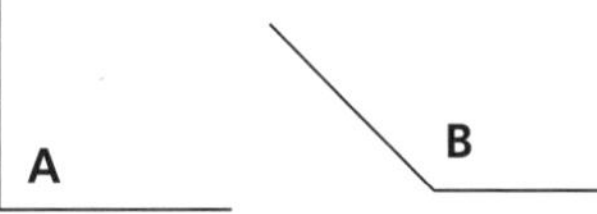

a Which angle is the largest? ______

b Which angle is the smallest? ______

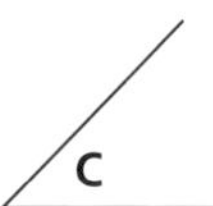

7. **a** Draw the axis of symmetry.

b What shape would I get if I fold this diamond along an axis of symmetry?

______

8:4 Extension ☐ out of 6

1. What is the greatest number of times that shape A could be cut from shape B? ______

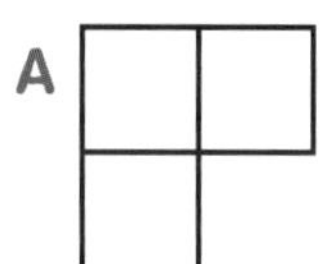

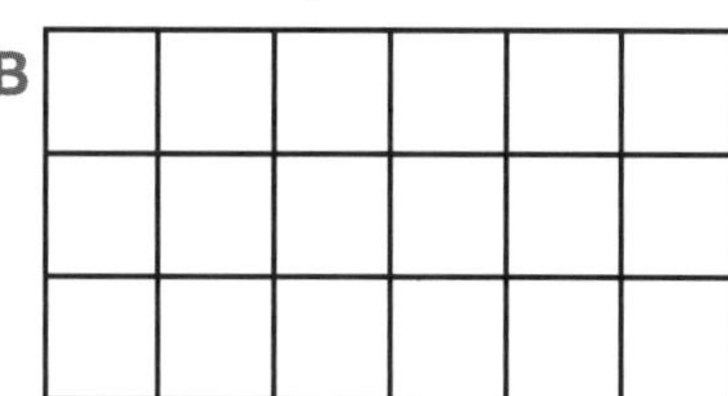

2. Draw a flip and slide of shape A.

Shape A

Flip of A

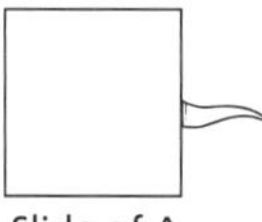
Slide of A

3. Regular shapes have all angles and all sides equal. What regular shape uses this angle?

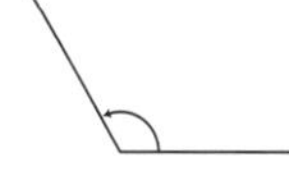

______

4. Which 2D shape has the greatest number of axes of symmetry? ______

5. Which shapes can be made with 4 right angles? ______

6. 24 + 16 + 12 + 28 ______

Challenge

Draw different 2D shapes that have 4 angles.

Concept

Some numbers are almost doubles.

Near doubling

6 + 7 = 6 + 6 + 1
= 12 + 1
= 13

a 8 + 9 = ______ **b** 16 + 15 = ______

c 7 + 8 = ______ **d** 24 + 23 = ______

e 20 + 21 = ______ **f** 35 + 36 = ______

g 32 + 33 = ______ **h** 13 + 14 = ______

i 16 + 17 = ______ **j** 26 + 25 = ______

9:1 out of 17

1. 5 × 2 ____
2. 5 × 4 ____
3. 8 × 2 ____
4. 8 × 4 ____
5. 63 + 25
6. 12 shared among 3 ____
7. \$5 + \$5 + \$5 ____
8. To 5 add 3. ____
9. Double 7. ____
10. 36 − 21

11. Write the mixed numeral for the shaded part.

a 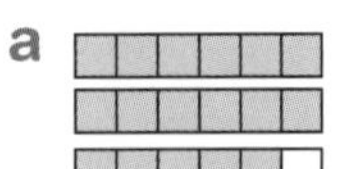 

b 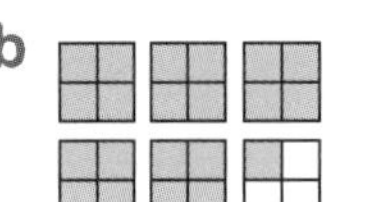 

12. 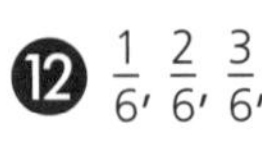 $\frac{1}{6}, \frac{2}{6}, \frac{3}{6},$ 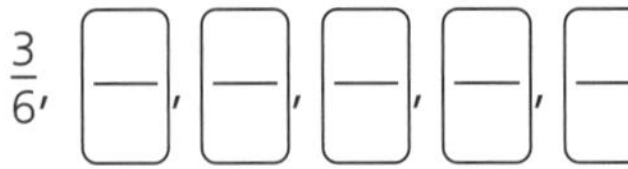

13. A B C

Which picture shows:

a a turn? ____ b a flip? ____

14. This angle is called a

r ____ a ____.

15. 30 000 + 9000 + 200 + 10 + 8 ____

16. Order these numbers from smallest to largest:
35 836 35 335 35 782

____

17. Draw 3 axes of symmetry on this circle.

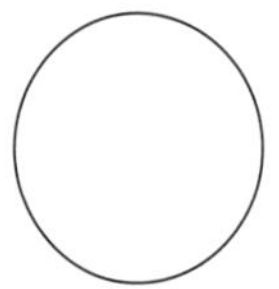

9:2 out of 16

1. 67 + 22 ____
2. 39 − 9 ____
3. 3 × 10 ____
4. 3 × 5 ____
5. 59 + 24
6. 9 groups of 5 ____
7. 9 times 10 ____
8. Double 25. ____
9. 35 subtract 7 ____
10. 63 + 28

11. How many axes of symmetry does this shape have? ____

12. Which angle is the largest? ____

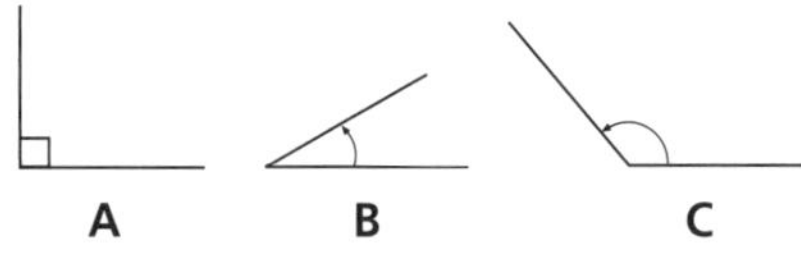

13. Is this a flip, a slide or a turn? ____

14. Complete this two-way table for these faces.

Smiley face	Circle	Square	Rhombus
Happy			
Sad			

15. 30 000 + 4000 + 20 + 9 ____

16. Draw 4 axes of symmetry.

Use place value to change these numbers.

a

Number	
45 673	+ 1000
	− 100
	+ 10
	− 1000

b

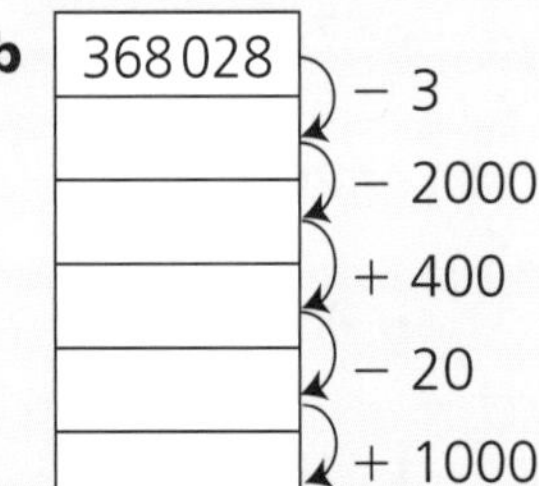

368 028	
	− 3
	− 2000
	+ 400
	− 20
	+ 1000

c

209 671	
	+ 200
	− 400
	− 2000
	− 40
	− 3000

Understanding place value makes this easy.

 ISBN 978 0 6557 0884 1

9:3 ☐ out of 10

1.

tens	ones
2	9
+3	1

2.

tens	ones
5	9
+1	2

3. List these angles in order of size from smallest to largest. ______

A 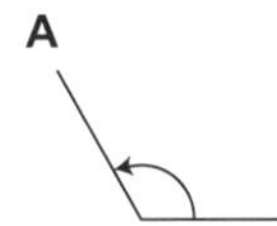 B  C

4. Draw a slide and a flip of this pattern.

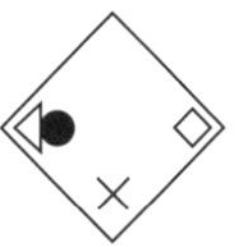

Slide

Flip

5. Draw the axes of symmetry of this shape.

6. 34 lions and 47 tigers. How many animals altogether? ______

7. How many bats? ______

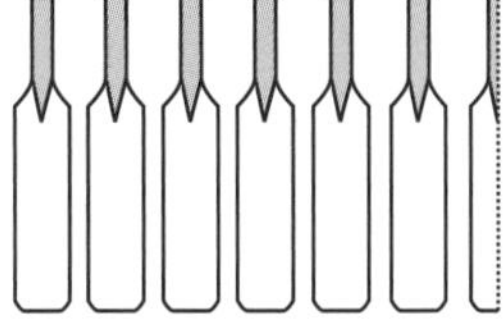

8. Order these numbers from smallest to largest:
79 463   79 640   79 465

______

9. 40 000 + 2000 + 700 + 30 + 9 ______

10. The value of the 4 in 354 982 is ______

9:4 Extension ☐ out of 7

1. Complete this tile pattern by flipping tiles to the right or down.

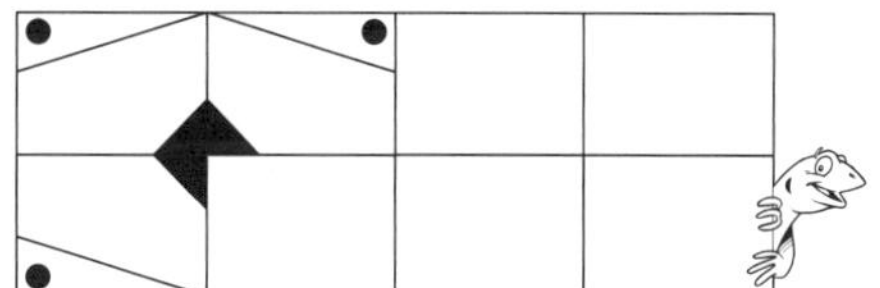

2. 34 frogs, 45 snakes and 36 lizards. How many creatures altogether? ______

3. 20 + 500 000 + 30 000 + 9 ______

4. 500 000 − 200 000 ______

5. 100 − 25 − 25 − 25 ______

6. How many different ways can you arrange these in a row?

______

7. This prism is 12 cm high. How high is:

a one slice? ______

b three slices? ______

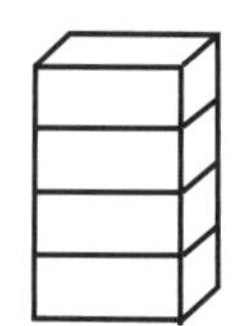

Challenge

What do you know about the number 838?

______

______

______

______

______

______

Use place value to change these numbers.

a

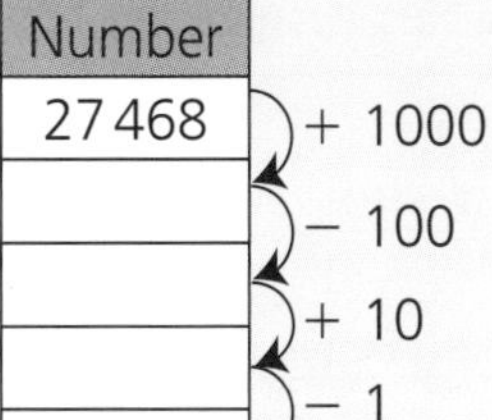

Number	
27 468	+ 1000
	− 100
	+ 10
	− 1

b

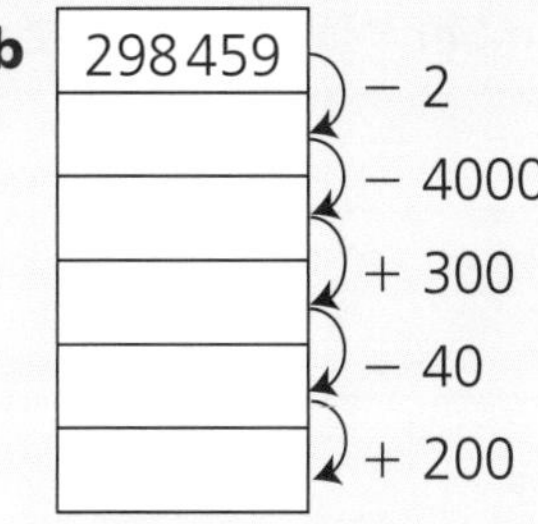

298 459	
	− 2
	− 4000
	+ 300
	− 40
	+ 200

c

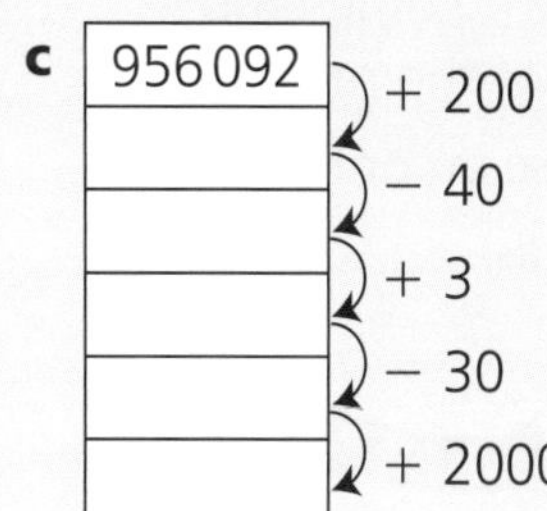

956 092	
	+ 200
	− 40
	+ 3
	− 30
	+ 2000

Use place value to make these easier.

10:1 ☐ out of 17

1. 16 − 7 ______
2. 13 − 8 ______
3. 15 − 6 ______
4. 21 − 4 ______
5. $\begin{array}{r} 23 \\ -15 \\ \hline \end{array}$
6. Double 6. ______
7. 15 + 13 − 13 ______
8. 84 m + 17 m ______
9. 6 multiplied by 10 ______
10. $\begin{array}{r} 37 \\ +38 \\ \hline \end{array}$

11. Order these numbers from smallest to largest:
91 708   92 087   92 807

______

12. 20 000 + 6000 + 100 + 80 + 4 ______

13. The value of the 9 in 264 982 is ______.
14. **a** Colour 2 out of 10 or $\frac{2}{10}$. **b** Circle $\frac{1}{5}$.

c Is $\frac{2}{10}$ the same as $\frac{1}{5}$? ______

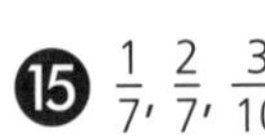

15. $\frac{1}{7}$, $\frac{2}{7}$, $\frac{3}{10}$, ☐, ☐, ☐, ☐

16. If a hat contains 2 red and 4 blue counters, how many counters would need to be drawn out, to be sure of getting a:

a red counter? ______

b blue counter? ______

17. 30 000 + 2000 + 70 + 1 ______

10:2 ☐ out of 17

1. 14 − 4 ______
2. 26 − 6 ______
3. 26 − 9 ______
4. 5 × 4 ______
5. $\begin{array}{r} 46 \\ +25 \\ \hline \end{array}$
6. 7 tens + 9 ones ______
7. 30c + 30c + 30c ______
8. 50 shared by 5 ______
9. 18 − 6 − 6 ______
10. $\begin{array}{r} 51 \\ -37 \\ \hline \end{array}$

11. Complete this two-way table for these shapes.

Shapes	Rhombus	Pentagon	Rectangle
Shaded			
Not shaded			

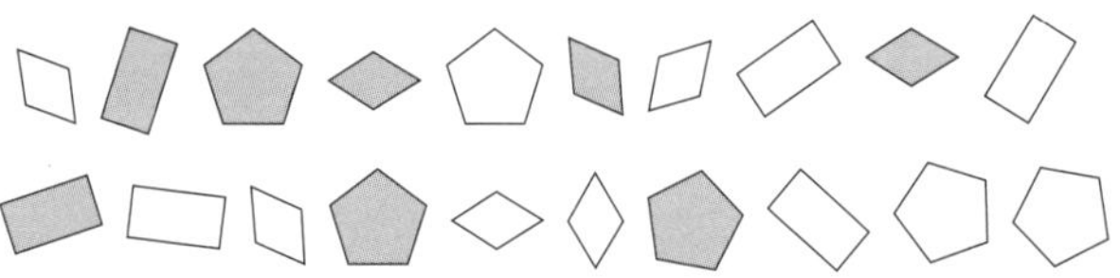

12. Order these numbers from smallest to largest:
28 463   28 098   28 439

______

13. True (T) or false (F)?

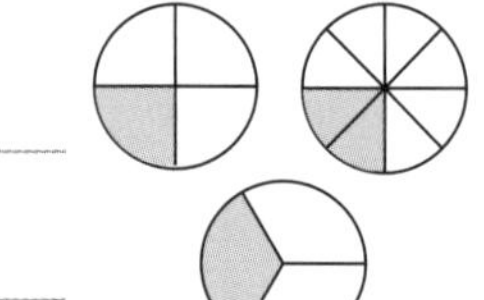

a $\frac{1}{4} = \frac{2}{8}$ ______

b $\frac{1}{4}$ is less than $\frac{1}{3}$. ______

14. The value of the 9 in 269 482 is ______.
15. 400 000 + 80 000 + 3000 + 400 + 3 ______
16. List all the posssible outcomes if two coins are tossed. ______
17. How many days in 2 weeks? ______

When you throw a dice, are all the numbers equally likely? ______

Throw a dice 40 times and keep a tally of the results. You could use an online dice to do this.

Number	1	2	3	4	5	6
Tally						

 AUSTRALIAN SIGNPOST MATHS 4 MENTALS • ISBN 978 0 6557 0884 1

10:3   ☐ out of 11

1.

tens	ones
3	9
+1	4

2.

tens	ones
2	7
+3	5

3. 10 000 + 5000 + 700 + 90 + 4 ______

4. The value of the 5 in 375 607 is ______.

5. If a right angle is a quarter turn, what is a straight angle? ______

6. $\frac{1}{9}, \frac{2}{9}, \frac{3}{9}$, ☐, ☐, ☐, ☐

7. Order these numbers from smallest to largest:
67 502   75 580   67 590
______

8. a How many 2s would you expect in 6 spins? ______
b How many 3s would you expect in 9 spins ? ______

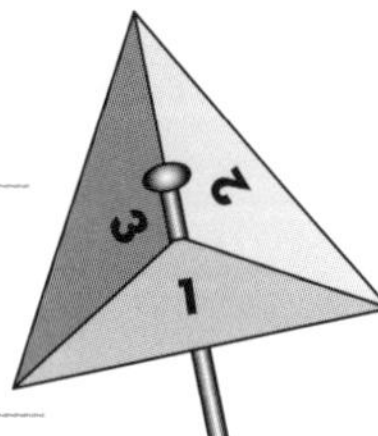

9. Write an equivalent fraction for:
a $\frac{1}{2}$ = ______   b $\frac{5}{5}$ = ______
c $\frac{4}{5}$ = ______   d $\frac{4}{8}$ = ______

10. 1006, 1007, 1008, ______, ______, ______

11. The numeral for fifty-six thousand, six hundred and nineteen. ______

10:4   Extension   ☐ out of 5

1. a 400 000 + 50 + 2000 + 10 000 ______
b 45 000 + 35 ______
c 32 000 + 5000 + 9 ______

2. Complete this picture if the broken line is an axis of symmetry.

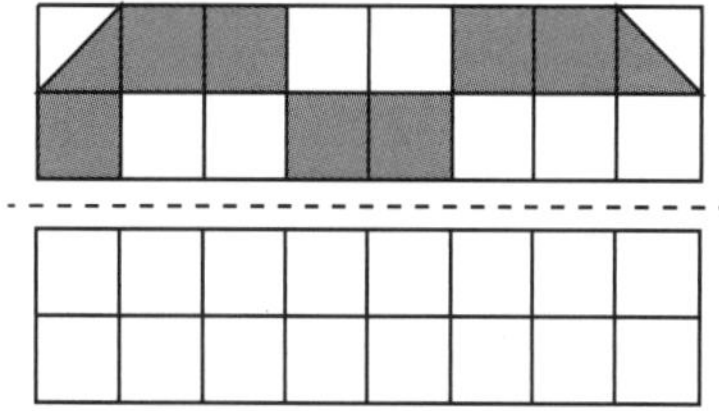

3. What fraction of $1 is:
a 50c? ☐   b 25c? ☐   c 75c? ☐

4. $\frac{1}{4}$ plus $\frac{1}{4}$.

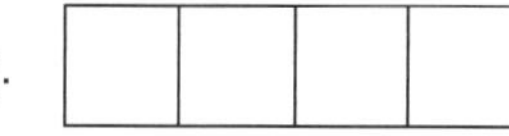

☐

5. Twenty people stood in a straight line holding hands.
How many hands were held? ______

Challenge

Draw and label fractions that are equivalent to $\frac{1}{2}$.

a

Part shaded = ____ out of 6
= ____ out of 2

b

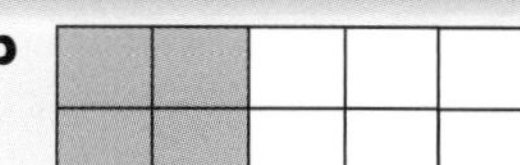

Part shaded = ____ out of 10
= ____ out of 5

3 sixths and 1 half are equal fractions.

c

Part shaded = ____ out of 8
= ____ out of 4

d

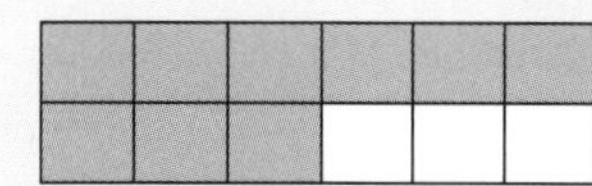

Part shaded = ____ out of 12
= ____ out of 4

 AUSTRALIAN SIGNPOST MATHS 4 MENTALS • ISBN 978 0 6557 0884 1

11:1 out of 16

1. 5 × 2 ____
2. 5 × 4 ____
3. 3 × 5 ____
4. 3 × 10 ____
5. 56 + 24
6. 18 divided by 2 ____
7. 24 minus 8 ____
8. Subtract 3 from 20. ____
9. Half of 14 ____
10. 51 − 33
11. True (T) or false (F)?
   a $\frac{1}{2} = \frac{1}{2}$ ____
   b $\frac{2}{4} = \frac{4}{8}$ ____

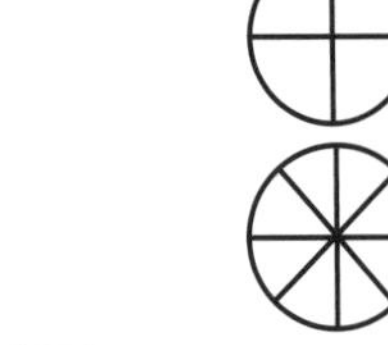

12. $\frac{1}{8}$, $\frac{2}{8}$, $\frac{3}{8}$, ☐, ☐, ☐, ☐, ☐

13. What temperature is shown for each?

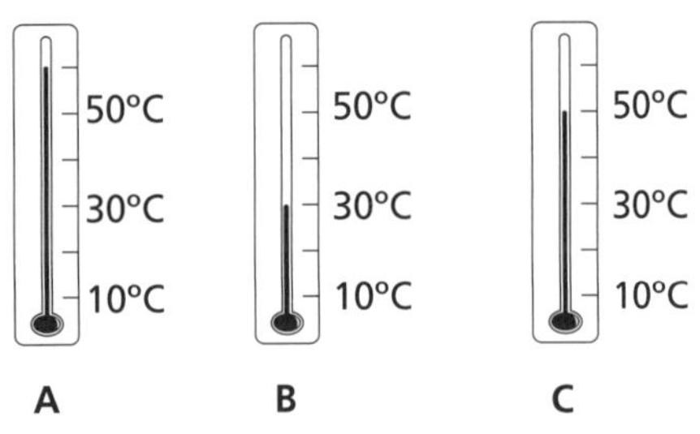

____ ____ ____

14. I have a 50c coin, a 20c coin and a 10c coin. How much money do I have? ____
15. 5, 10, ____, ____, ____, 30, ____, ____
16. Draw circles to find 4 groups of 3. 4 × 3 = ____

11:2 out of 19

1. 7 × 2 ____
2. 7 × 4 ____
3. 8 × 5 ____
4. 8 × 10 ____
5. 37 + 45
6. 32 + 16 − 16 ____
7. 12 divided by 4 ____
8. 16 ÷ 2 ____
9. 16 ÷ 4 ____
10. 63 − 39
11. Complete:
   a $\frac{6}{8} = \frac{\square}{4}$
   b $\frac{1}{3} = \frac{2}{\square}$
12. 0, $\frac{1}{4}$, $\frac{2}{4}$, ____, 1, ____, ____
13.

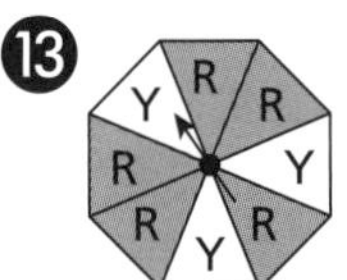

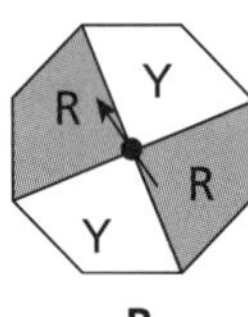

red (R)
yellow (Y)

Which spinner is more likely to land on red than yellow? ____

14. Does water freeze at 0°C? ____
15. Write fifty-three degrees Celsius in short form. ____
16. 1, 3, 5, ____, ____, ____, ____, ____
17. a Does 20 + 18 = 40 + 8? ____
   b Does 30 + 16 = 40 + 6? ____
18. a 43 + 9 = 40 + ____ b 27 + 7 = 30 + ____
   c 35 + 8 = 40 + ____ d 64 + 9 = 70 + ____
19. 600 000 + 40 000 + 2000 + 300 + 9 ____

What events have a 50–50 (or even) chance of happening?
Write down as many as you can.

____
____
____
____
____

 • *AUSTRALIAN SIGNPOST MATHS 4 MENTALS* • ISBN 978 0 6557 0884 1

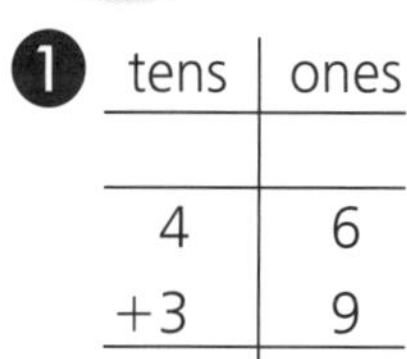

11:3 ☐ out of 6

1

tens	ones
4	6
+3	9

2

tens	ones
5	7
+1	7

3 True (T) or false (F)?

a $\frac{4}{4} = 1$ ______

b $\frac{1}{3}$ is less than $\frac{1}{4}$. ______

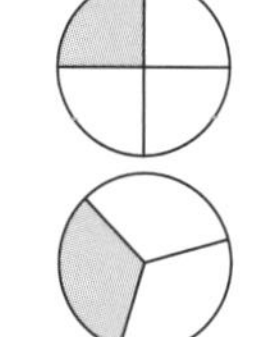

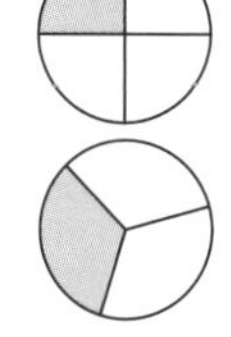

4 $2\frac{1}{2}$, 3, ______, 4, ______, ______

5

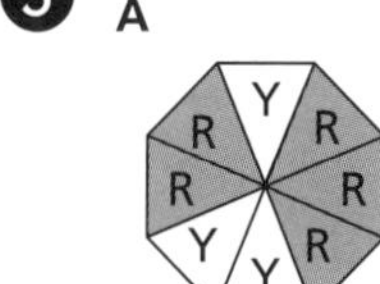

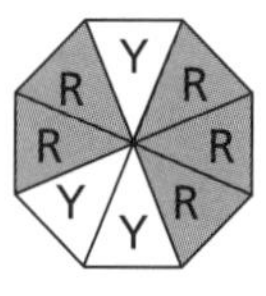

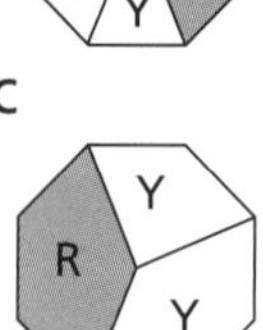

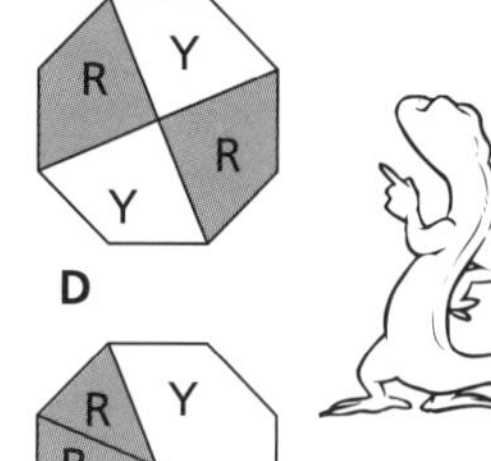

a Which spinner is more likely to land on R than Y? ______

b Which spinner is as likely to land on R than Y? ______

6 Colour the mercury to match the temperature.

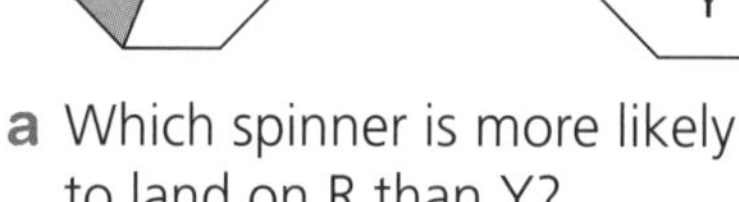

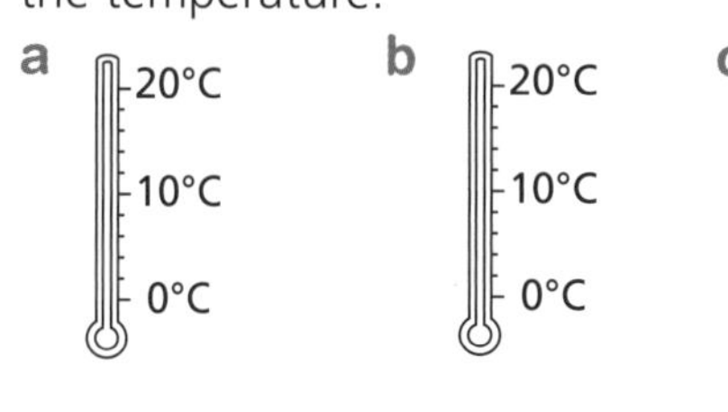

a 12°C **b** 6°C **c** 19°C

11:4 Extension ☐ out of 5

1

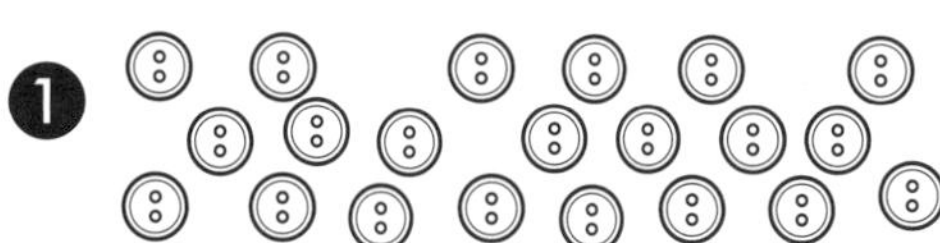

$\frac{1}{3}$ means 1 share out of 3.

What is $\frac{1}{3}$ of 21? ______

2 What fraction of $1 is:

a 30c? ______

b 75c? ______

3 If this shape is turned, which shape below could it *not* look like?

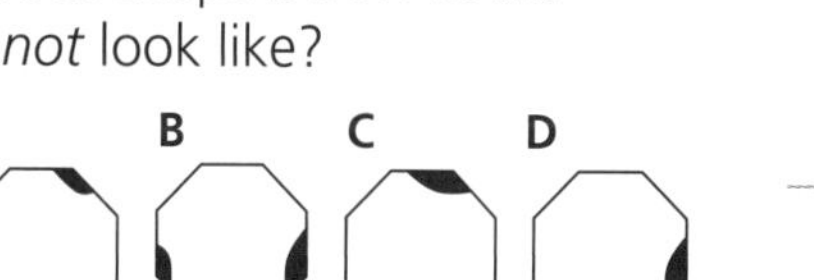

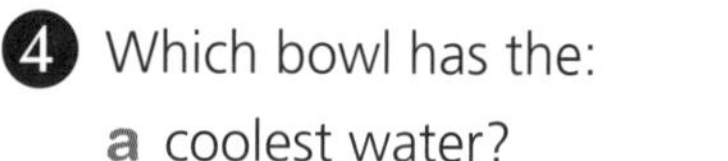

A B C D ______

4 Which bowl has the:

a coolest water? ______

b warmest water? ______

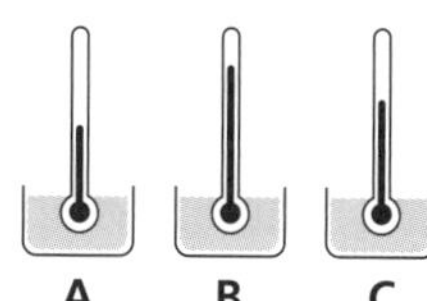

5 45, 47, 49, ______, ______, ______, ______, ______

Challenge

Complete each number sentence so there is an equation on each side of the = sign.

a 28 + 9 = 30 + ______ **b** 37 + 6 = 40 + ______

c 45 + 9 = 50 + ______ **d** 56 + 8 = 60 + ______

e 84 + 8 = 90 + ______ **f** 76 + 7 = 80 + ______

g 57 + 6 = ______ **h** 49 + 6 = ______

i 47 + 7 = ______ **j** 67 + 6 = ______

12:1

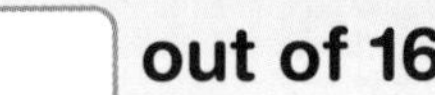

out of 18

1. 15 + 10 ______
2. 3 × 5 ______
3. 10 ÷ 2 ______
4. 23 − 4 ______
5. $\begin{array}{r} 37 \\ +38 \\ \hline \end{array}$
6. Take 14 from 20. ______
7. Add 15 and 6. ______
8. Double 8. ______
9. Half of 12 ______
10. $\begin{array}{r} 64 \\ -38 \\ \hline \end{array}$
11. Write seventeen degrees Celsius in short form. ______
12. Write the number after 43 563. ______
13. 47 + 35 ______

    47

14. Write the number two hundred and thirty-one thousand, five hundred and eight. ______
15. Round 4673 to the nearest:
    a thousand ______ b ten ______
16. Draw the line of symmetry on this shape.

17.

What part is shaded? ______ out of ______

18. Which shape has the largest area? ______

A

B

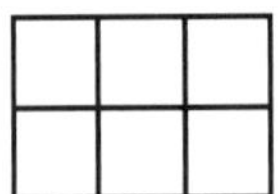

12:2

out of 16

1. 19 + 13 ______
2. 15 + 17 ______
3. 48 + 7 ______
4. 6 × 10 ______
5. $\begin{array}{r} 58 \\ +26 \\ \hline \end{array}$
6. 37 + 24 − 24 ______
7. Half of 20 ______
8. Double 9. ______
9. 20 + 8 − 3 ______
10. $\begin{array}{r} 72 \\ -57 \\ \hline \end{array}$

11. Does water boil at 100°C? ______
12. a 25 + 8 = 30 + ______
    b 37 + 9 = 40 + ______
    c 37 + 7 = 40 + ______
    d 68 + 7 = 70 + ______
13. Write 4093 in words.

    ______
14. Round 7 835 465 to the nearest:
    a million ______
    b thousand ______
15. Write in order from largest to smallest.
    7 645 890 7 655 982 7 689 023

    ______
16. a 38 + 27 ______

    38

    b 53 − 28 ______

    53

Follow the instructions to flip, slide or turn these shapes.

slide flip turn flip turn slide flip flip turn

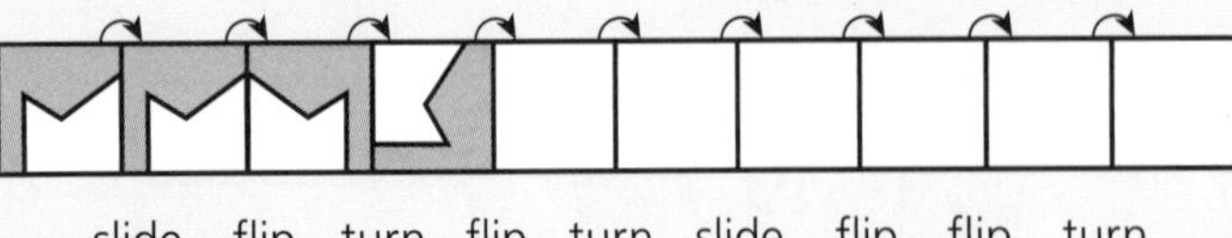

slide flip turn flip turn slide flip flip turn

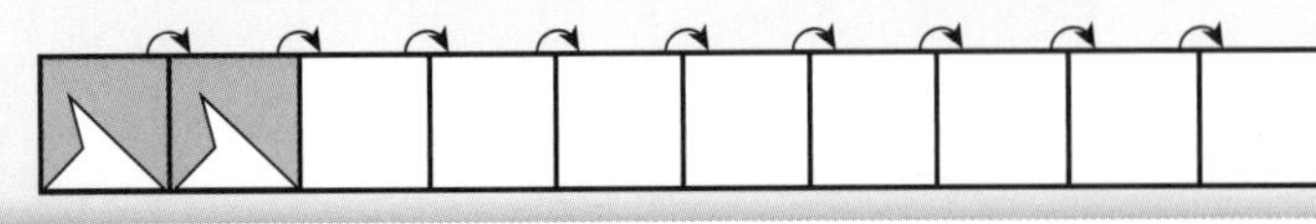

12:3 out of 9

1.

tens	ones
6	7
+ 4	5

2.

tens	ones
3	9
+ 6	3

3. a 29 + 6 = 30 + ____ b 45 + 7 = 50 + ____

   c 38 + 4 = 40 + ____ d 56 + 9 = 60 + ____

4. Write the numeral for:
   300 000 + 20 000 + 5000 + 400 + 90 + 2 ________

5. Write in order from largest to smallest.
   3 264 879 3 268 980 3 269 671

   ____

6. Write the value of the 7 in 9 672 536. ________

7. Round 5 643 892 to the nearest:
   a million ________
   b thousand ________

8. Here is a h ____________
   s ____________.
   There are ________ squares in one row.

9. a 47 + 39 ________

47

   b 41 − 28 ________

41

12:4 out of 10 Extension

1. 1, 2, 4, 8, ____, ____, ____, ____
2. 97 + 44 ________

   97

3. Which is larger: 4 × 6 or 5 × 5? ________
4. 50 − 20 − 15 − 10 ________
5. Number of faces plus the number of corners minus the number of edges.

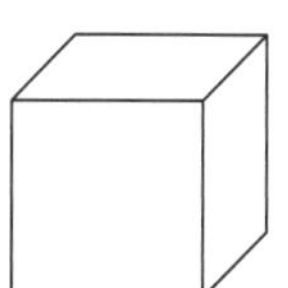

   ____

6. If 26 + 28 = 54, then:

   a 54 − 28 = ______ b 54 − 26 = ______

7. a 30 minutes after 7:15.

   ____

   b 30 minutes before 7:15.

   ____

8. 80 − 25 − 25 − 27 ________
9. 50 + 300 + 80 000 + 900 000 + 6000 ________
10. How many pairs in 24 shoes? ________

Challenge

Make a number pattern using the rule:

a Add 25. ________

b Subtract 30. ________

c Add 9. ________

d Subtract 6. ________

e Add 50. ________

Use place value to change these numbers.

a

Number	
57 648	+ 100
	− 1000
	+ 1
	− 10

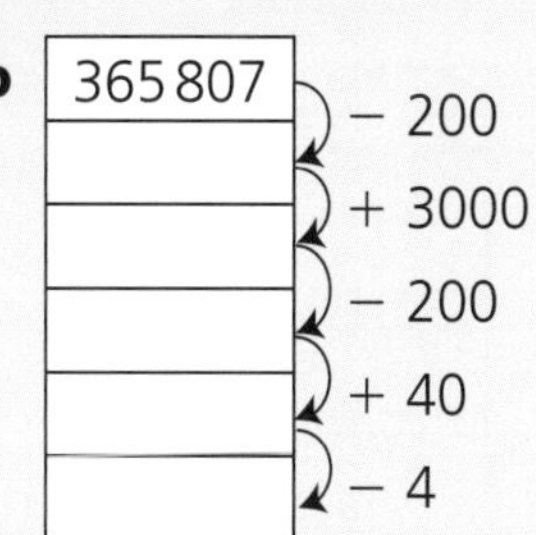

b

365 807	
	− 200
	+ 3000
	− 200
	+ 40
	− 4

c

473 896	
	+ 100
	− 20
	+ 3
	− 20
	+ 3000

13:1 ☐ out of 16

1. 1 × 4 ______
2. 1 × 8 ______
3. 2 × 4 ______
4. 2 × 8 ______
5. $\begin{array}{r} 41 \\ -24 \\ \hline \end{array}$
6. 10 groups of 8 ______
7. 0 rows of 8 ______
8. 2 × 8 + 8 ______
9. 3 × 8 ______
10. $\begin{array}{r} 36 \\ +45 \\ \hline \end{array}$
11. Write in order from largest to smallest.
    578 099   578 354   578 936
    ____
12. Round 37 528 to the nearest:
    a hundred ______
    b thousand ______
13. a What part is shaded?
    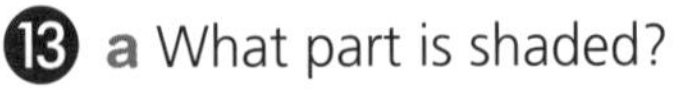
    ______
    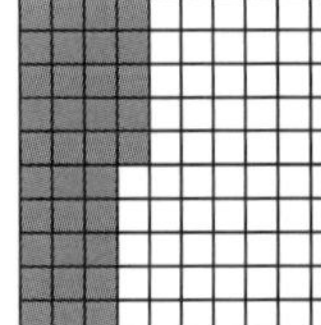
    b What part is not shaded?
    ______
14. a 4, 8, 12, ____, ____, ____, ____, ____
    b 3, 6, 9, ____, ____, ____, ____, ____
    c 6, 12, 18, ____, ____, ____, ____, ____
    d 8, 16, 24, ____, ____, ____, ____, ____
15. How many legs are on 2 spiders? ______

16. Draw 3 pentagons.
    How many sides altogether? ______

13:2 ☐ out of 15

1. 4 × 4 ______
2. 4 × 8 ______
3. 5 × 4 ______
4. 5 × 8 ______
5. $\begin{array}{r} 82 \\ -37 \\ \hline \end{array}$
6. Double 6 × 4 ______
7. 6 × 8 ______
8. 9 multiplied by 4 ______
9. 8 times 4 ______
10. $\begin{array}{r} 57 \\ +26 \\ \hline \end{array}$

11. Round 4 627 290 to the nearest:
    a million ______
    b thousand ______
12. a 56 + 37 ______
    56
    b 73 − 26 ______

    73
13. Write in order from largest to smallest.
    8 356 809   8 356 263   8 356 978
    ____
14. Write the value of the 5 in 7 465 978. ______
15. a Shade 80 out of 100.
    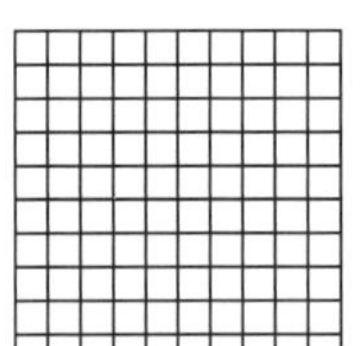

    b What part is not shaded?
    ______ out of ______ or 0·______.

×	5	10	4	0	2	6	1	3
2								
4								
8								
5								
10								

 ISBN 978 0 6557 0884 1

13:3 out of 9

1

tens	ones
4	0
−	7

2

tens	ones
5	0
−3	4

3 Round 284 472 to the nearest:

a hundred ______

b thousand ______

4 a 68 + 24 ______

b 62 − 48 ______

5 The fraction shown is ______ hundredths or 0·______.

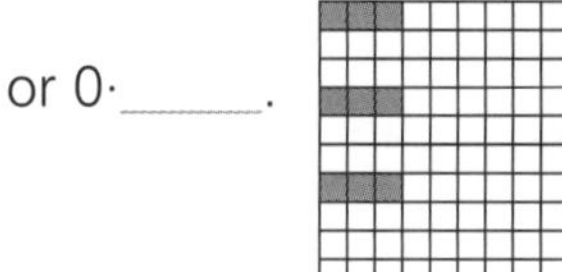

6 a 4, 8 ,12, ______, ______, ______, ______

b 8, 16 , 24, ______, ______, ______, ______

7 How many legs are on:

a 5 octopuses? ______

b 10 octopuses? ______

c 6 octopuses? ______

8 600 000 + 40 000 + 1000 + 800 + 50 + 8 = ______

9 Order these numbers from largest to smallest.
294 658 294 599 294 673

______

13:4 Extension out of 8

1 99 + 32 ______

2 a 60 + ______ = 100

b 100 − 40 = ______

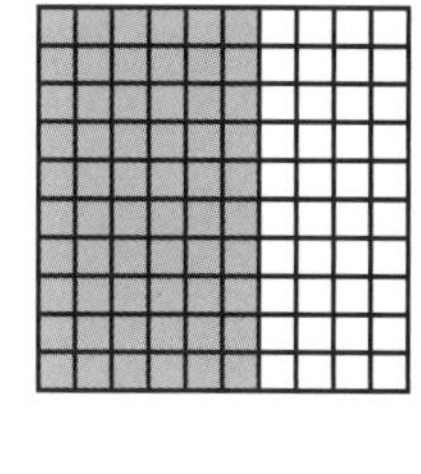

3 a 60 − □ = 32

b □ − 17 = 24

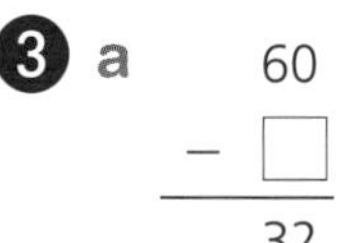

4

How many legs on 3 dogs, 3 ducks and 5 croodiles? ______

5 2 metres minus 50 cm. ______

6 a 100, 96 , 92, ______, ______, ______, ______

b 100, 92 , 84, ______, ______, ______, ______

7 How many sides on 6 octagons and 3 hexagons. ______

8 45, 47, 49, ______, ______, ______, ______, ______

Challenge

Complete each number sentence so there is an equation on each side of the = sign.

a 25 + 3 = 20 + ____ b 54 + 5 = 50 + ____

c 37 + 9 = 40 + ____ d 57 + 8 = 60 + ____

e 46 + 6 = ______ f 79 + 5 = ______

g 88 + 8 = ______ h 76 + 12 = ______

The jump strategy

Use a number line to jump the tens then the ones.

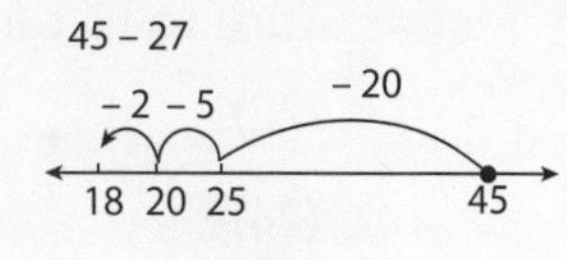

a 61 − 38 = ______ b 56 − 27 = ______

c 73 − 37 = ______ d 82 − 55 = ______

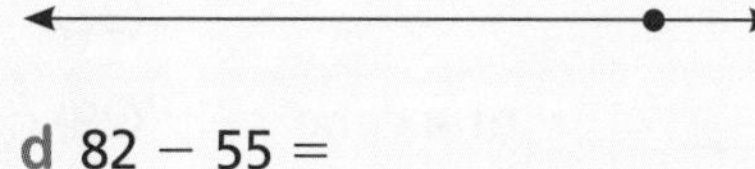

 • *AUSTRALIAN SIGNPOST MATHS 4 MENTALS* • ISBN 978 0 6557 0884 1

14:1 ☐ out of 16

1. 17 + 5 ______
2. 19 − 8 ______
3. 2 × 8 ______
4. 4 × 8 ______
5. $\begin{array}{r} 68 \\ +28 \\ \hline \end{array}$
6. Half of 16 ______
7. Double 5. ______
8. 5 multiplied by 8 ______
9. 10 times 8 ______
10. $\begin{array}{r} 53 \\ -37 \\ \hline \end{array}$

11. Write the fraction for the part shaded.

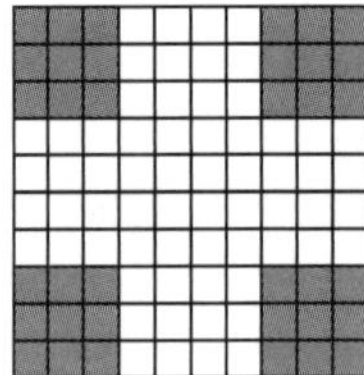

______ out of ______

or 0· ______.

12. **a** 0·4, 0·5, 0·6, ______, ______, ______, ______
    b 0·35, 0·36, 0·37, ______, ______, ______

13.

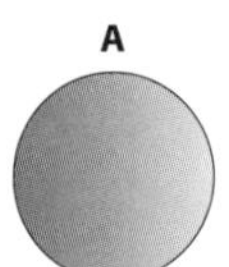

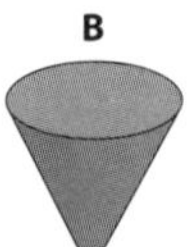

a Which shape is a cone? ______
b Which shape has 2 flat surfaces? ______

14. **a** Write the decimal for 5 tenths. ______
    b Write the decimal for $\frac{3}{10}$. ______

15. In one tank there are 27 fish and in another there are 12.

How many fish are there altogether? ______

16. Draw an octagon. It has ______ sides.

14:2 ☐ out of 20

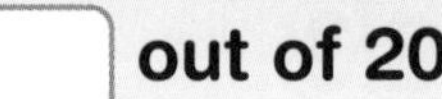

1. 65 − 10 ______
2. 4 × 8 ______
3. 8 × 8 ______
4. 7 × 8 ______
5. $\begin{array}{r} 53 \\ +39 \\ \hline \end{array}$
6. 8 × 8 + 8 ______
7. 9 × 8 ______
8. 36 + ______ = 40
9. 54 + ______ = 60
10. $\begin{array}{r} 74 \\ -38 \\ \hline \end{array}$

11. **a** Write the decimal for 9 tenths. ______
    b Write the decimal for $4\frac{6}{10}$. ______

12. On this hundred square, the part covered is:

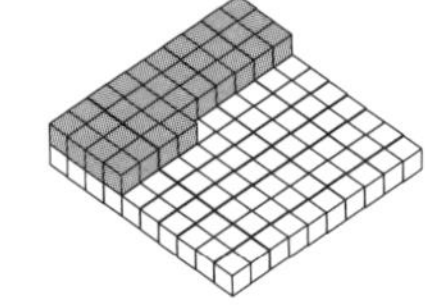

______ ______

or 0· ______.

13. Which decimal is larger, 0·42 or 0·37? ______

14. $\frac{3}{10}$, $\frac{4}{10}$, $\frac{5}{10}$, ☐, ☐, ☐, ☐

15. Is $\frac{18}{18}$ equal to 1? ______

16. What two shapes are used to make this model?

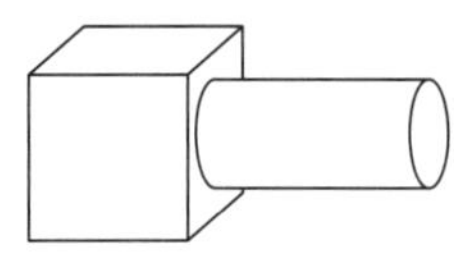

______
______

17. How many minutes from 9:30 to 9:45? ______
18. 30 + 40 + 60 + 70 ______
19. Is 356 closer to 300 or 400? ______
20. I was given 5 cards on Monday, 13 on Tuesday and 15 on Wednesday. How many cards was I given? ______

Turn to ID card C on page 8. Give the answers for these numbers.

Answer only the numbers shown.

(22) ______ (23) ______

(24) ______ (25) ______

(26) ______ (27) ______

(28) ______ of a cube (29) net of a ______

(30) net of a ______

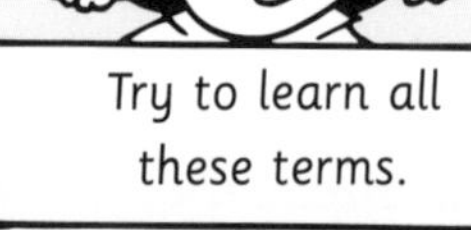

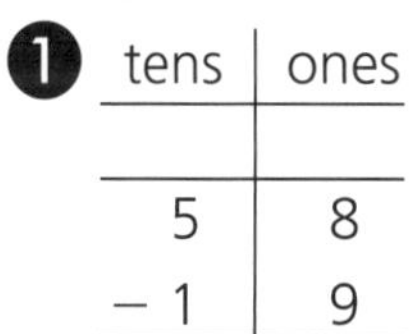

out of 11

1
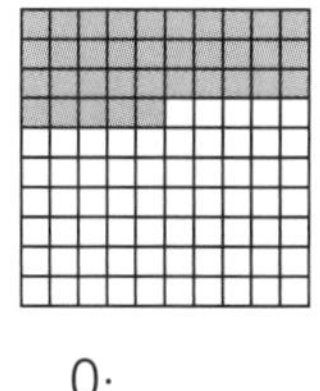

tens	ones
5	8
− 1	9

2
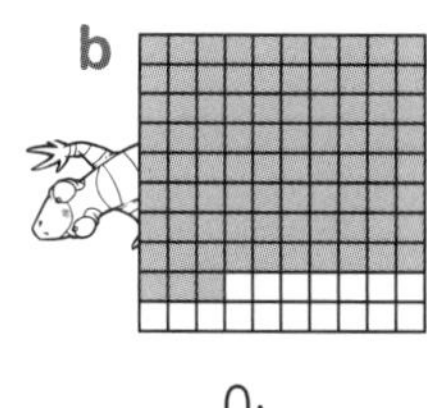

tens	ones
\$3	7
−\$1	9

3 What decimal is shown?

a 0· ______

b 0· ______

4 a $\frac{3}{10}$, $\frac{4}{10}$, $\frac{5}{10}$, ☐, ☐, ☐, ☐, ☐, ☐

b 0·75, 0·76 , 0·77, ______, ______, ______

5 This shape is a ______________.
It has ____ round surface.

6

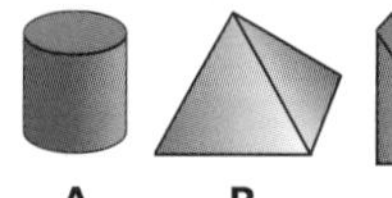
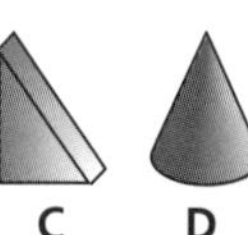

A B C D

a Which of these objects is a pyramid? ______

b Which of these objects is a prism? ______

7 42 stamps plus 36 stamps. ______

8 Write the numeral for
six thousand and twelve. ______

9 50c + 20c + 20c + 5c ______

10 The change from \$2.00, if I buy a banana for 80c. ______

11 Jim had 63 stamps. He gave 10 to his sister. How many did he have left? ______

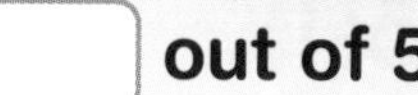

out of 5

1 13 hundredths plus 21 hundredths.

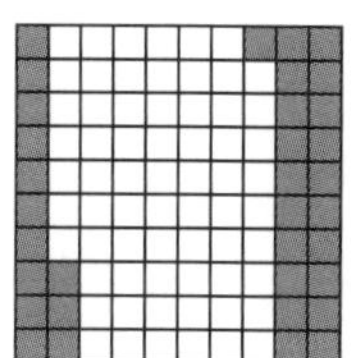

____

or 0· ______.

2 8 mugs are in each box.
How many mugs would be in 12 boxes? ______

3 a The number of faces. ______

b The number of faces + corners − edges.
______ + ______ − ______ = ______

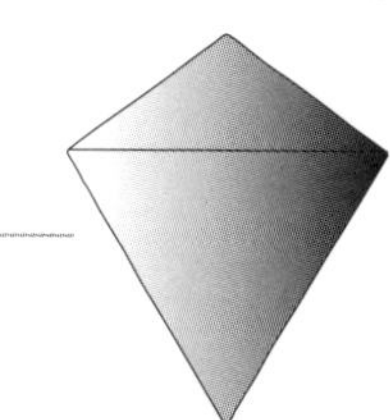

4 If two slices of bread make 4 sandwiches, how many sandwiches can be made from a loaf of 24 slices?

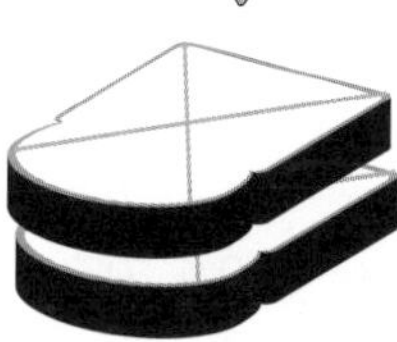

____

5 a 98, 89, 80, ______, ______, ______, ______

b 87, 79, 71, ______, ______, ______, ______

Challenge

Draw a pyramid. Label and describe its features.

____

____

____

____

____

	2	4	6	8	10
× 4					

	1	2	3	4	5
× 8					

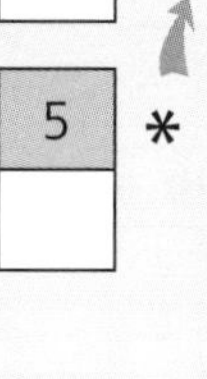

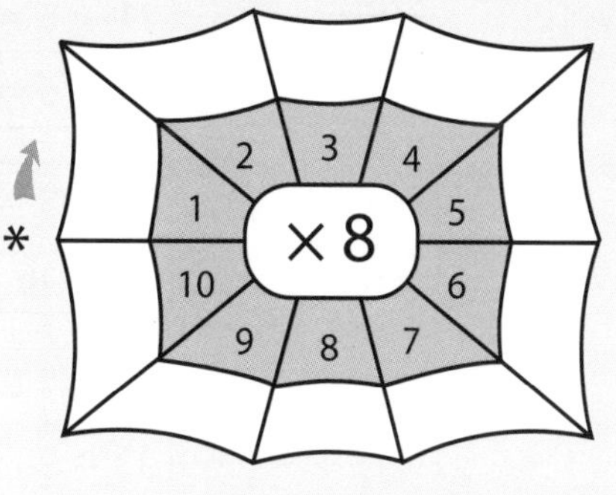

 • *AUSTRALIAN SIGNPOST MATHS 4 MENTALS* • ISBN 978 0 6557 0884 1

15:1

☐ out of 16

1. 3 × 8 ______
2. 10 × 8 ______
3. 14 + 14 ______
4. 27 − 12 ______
5.
```
  5 9
+ 3 4
```
6. Double 13. ______
7. 5 multiplied by 8 ______
8. Half of 40 ______
9. 35 minus 23 ______
10.
```
  5 3
− 3 5
```

11. What decimal is shown?

a 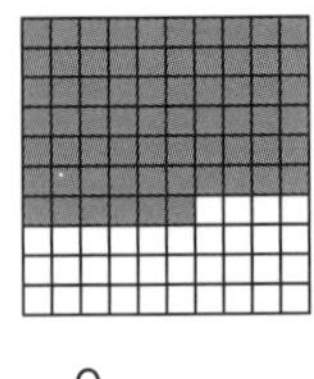

0· ______

b 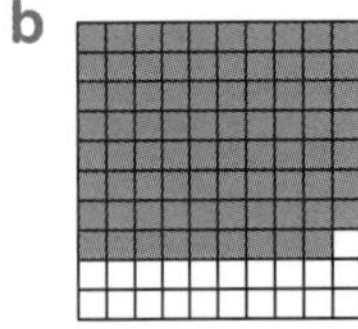

0· ______

12. a 0·1, 0·2, 0·3, ______, ______, ______, ______

    b 0·28, 0·29, 0·30, ______, ______, ______

13. True (T) or false (F)?

    a 13 hundreds = 0·13 ______

    b 5 hundredths = 0·05 ______

14. How many:

    a faces? ______

    b corners? ______

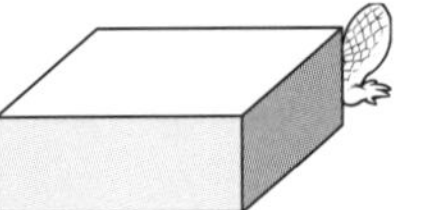

15. The total value of these coins in decimal form. ______

16. How many tens in 6982? ______

15:2

☐ out of 16

1. 29 + 15 ______
2. 35 − 25 ______
3. 40 − 24 ______
4. 4 × 8 ______
5.
```
  6 8
+ 2 5
```
6. Double 23. ______
7. Half of 46 ______
8. 6 times 8 ______
9. 9 multiplied by 8 ______
10.
```
  7 1
− 4 8
```

11. Which decimal is larger, 0·53 or 0·49? ______

12. Write the decimal modelled.

Ones	Tenths	Hundredths
0 •		

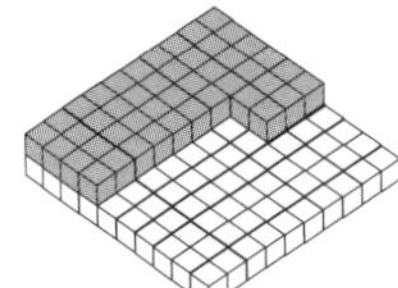

13. Write the value of the 5 in 9·53. ______

14. A B C D

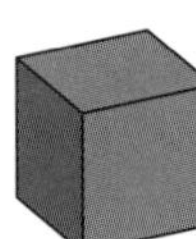

    a Which shape has only flat surfaces? ______

    b Which shape has 8 corners? ______

    c Which shapes have 1 curved surface? ______

15. Write the number that has:

    a 4 hundreds, 4 tens, 9 units, 6 tenths and 8 hundredths. ______

    b 7 hundreds, 2 tens, 7 units, 9 tenths and 4 hundredths. ______

16. Jason drew 2 monsters. He gave each one 18 legs. How many legs were there altogether? ______

Addition linked to subtraction

If 6 + 7 = 13, then 13 − 6 = 7, 13 − 7 = 6

a 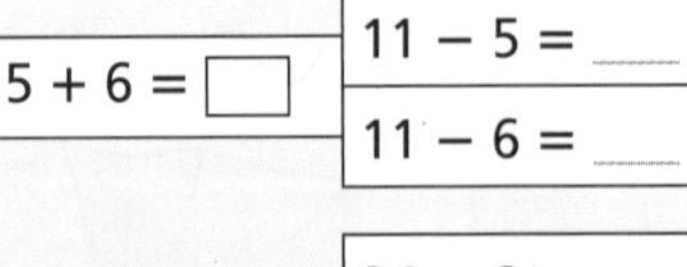

5 + 6 = ☐ | 11 − 5 = ___ | 11 − 6 = ___

b 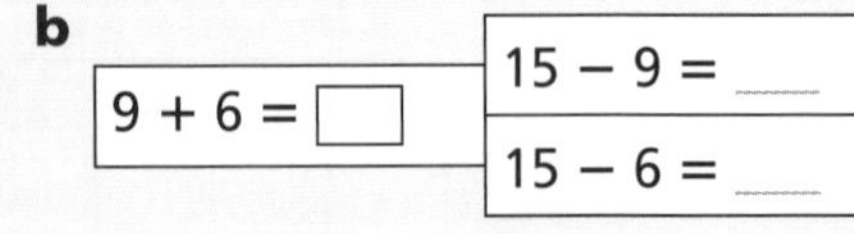

9 + 6 = ☐ | 15 − 9 = ___ | 15 − 6 = ___

c 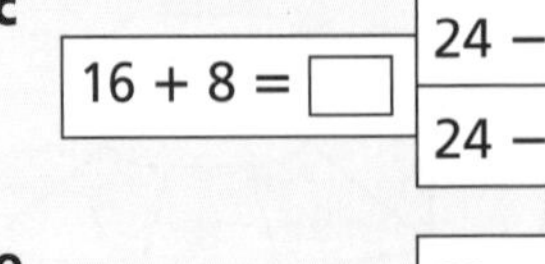

16 + 8 = ☐ | 24 − 8 = ___ | 24 − 16 = ___

d 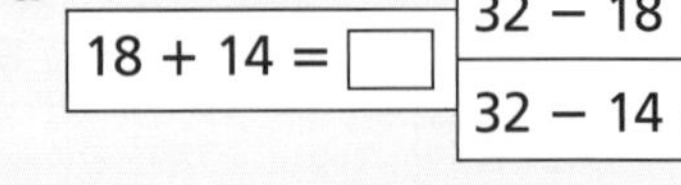

18 + 14 = ☐ | 32 − 18 = ___ | 32 − 14 = ___

e 28 + 5 = ☐ | 33 − 5 = ___ | 33 − 28 = ___

f 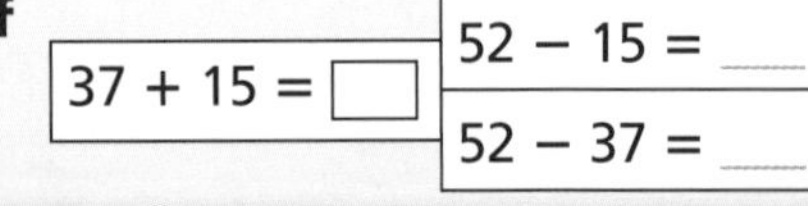

37 + 15 = ☐ | 52 − 15 = ___ | 52 − 37 = ___

 ISBN 978 0 6557 0884 1

15:3 ☐ out of 10

1.

tens	ones
2	2
− 1	3

2.

tens	ones
$3	5
−$1	7

3. Write the fraction for the part shaded. ______ out of ______

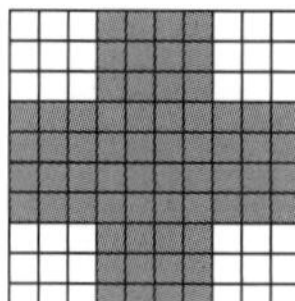

4. Write the decimal for:
   a zero point three seven ______
   b 8 tenths ______
5. Which decimal is larger, 0·2 or 0·6? ______
6. Which of these is not a prism? ______

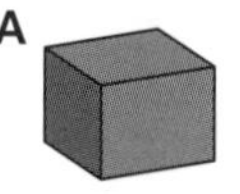

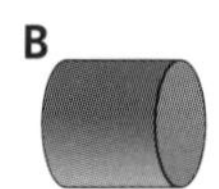

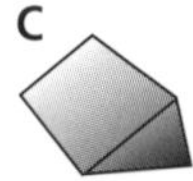

7. Sue's team scored 13 goals in a game of soccer. If they scored 4 goals in the first half, how many goals did they score in the second half? ______
8. A mouse was balanced by 6 coins. A book was balanced by 12 coins.

   How many mice would balance the book? ______
9. In a cupboard there were 18 cups, 10 bowls, 16 saucers and 13 plates. How many more cups were there than plates? ______
10. 15 + 15 + 15 + 15 ______

15:4 Extension ☐ out of 7

1. 13 + 13 + 13 + 13 ______
2. a 83 − 22 = 81 − ______ = ______
   b 68 − 34 = 64 − ______ = ______
3. I have 5 cubes.
   a How many faces? ______
   b How many corners? ______
   c How many edges? ______
4. How many three-digit numbers can be made using these three digits?
   a 8, 7 and 0 ______ b 5, 3 and 1 ______
5. 28 soldiers stood in each row. How many soldiers in three rows? ______
6. Tomorrow is Friday. What day was it 8 days ago? ______
7. Sixteen people stood in a straight line holding hands. How many hands were held? ______

Challenge

Draw a 3D object and describe it.

______

______

______

______

______

______

______

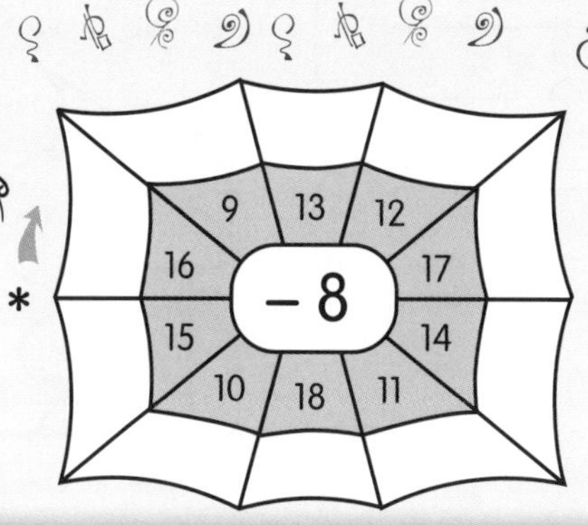

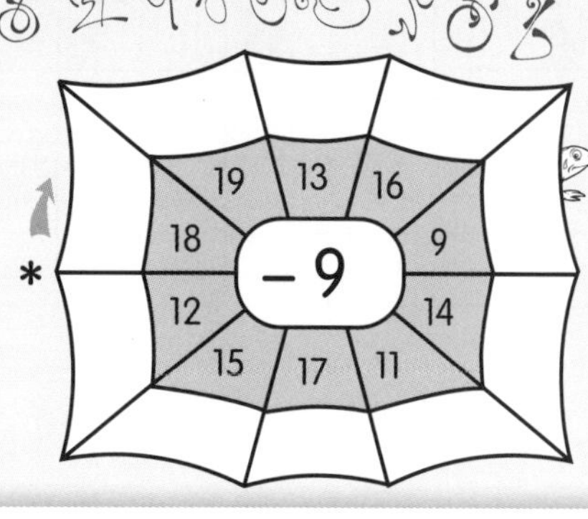

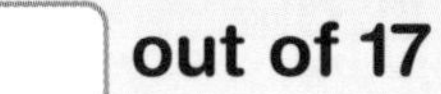

16:1 ☐ out of 16

1. 35 + 10 ______
2. 28 + 20 ______
3. 45 − 10 ______
4. 52 − 20 ______
5. $\begin{array}{r} 52 \\ +63 \\ \hline \end{array}$
6. \$4 + \$4 + \$4 ______
7. 37 subtract 20 ______
8. Digits in 70 746 ______
9. 37, 39, 41, ______, ______
10. $\begin{array}{r} 72 \\ +64 \\ \hline \end{array}$
11. Which decimal is larger, 0·78 or 0·72? ______
12. Write the decimal for zero point two three. ______
13. Write the decimal shown by this model. ______

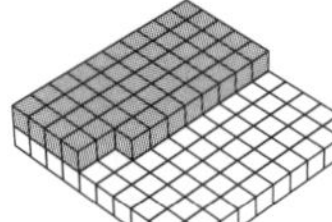

14. True (T) or false (F)?
    a 8 hundredths = 0·80 ______
    b 8 hundredths = 0·08 ______
15. What two objects fit together to make this shape?

______

16. Students owning birds

8, 4, 0

Budgies | Finches | Parrots | Cockies

    a How many own parrots? ______
    b How many students own parrots or cockies together? ______
    c How many students altogether? ______

16:2 ☐ out of 17

1. 61 + 34 ______
2. 85 − 24 ______
3. 57 − 35 ______
4. 53 + 35 ______
5. $\begin{array}{r} 45 \\ +82 \\ \hline \end{array}$
6. 1 m − 45 cm ______
7. \$5.60 − \$1 ______
8. 4 times 10 kg ______
9. Multiply 4 by 8 ______
10. $\begin{array}{r} 65 \\ +93 \\ \hline \end{array}$
11. Shade 65 out of 100. What part is not shaded? ______

12. Write the decimal shown here. ______

13. Write the number that has 4 hundreds, 4 tens, 9 units, 6 tenths and 8 hundredths. ______
14. How many colours are needed to colour a cube if adjoining faces cannot have the same colour? ______

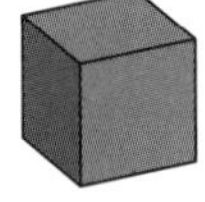

15. Sarah scored 22 goals and Josie scored 12. How many goals did they score? ______
16. One share is 10. How many is 7 shares? ______
17. List the value of the notes and coins needed to make \$231.25 ______

______

	2	4	6	8	10
× 3					

	1	2	3	4	5
× 6					

	7	5	9	3	1
× 3					

	8	10	7	9	6
× 6					

16:3 ☐ out of 8

1

tens	ones
$1	1
$5	7
+ $2	8

2

tens	ones
$2	3
$3	2
+$1	7

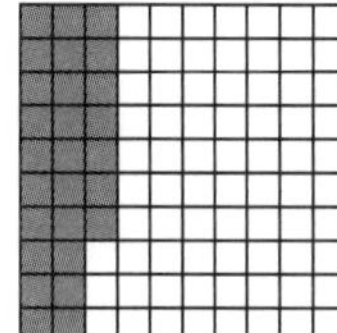

3 What fraction is shaded?

a ______ hundredths

b As a decimal this is

______.

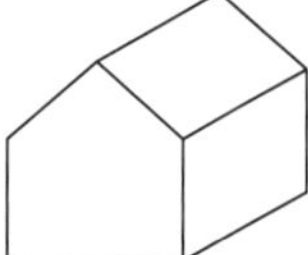

4 Is this a prism or a pyramid or neither?

______

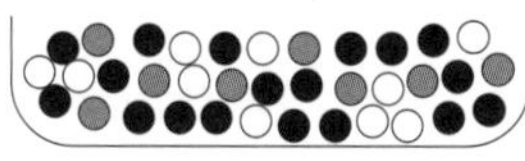

5 Complete the bar graph for this group of marbles.

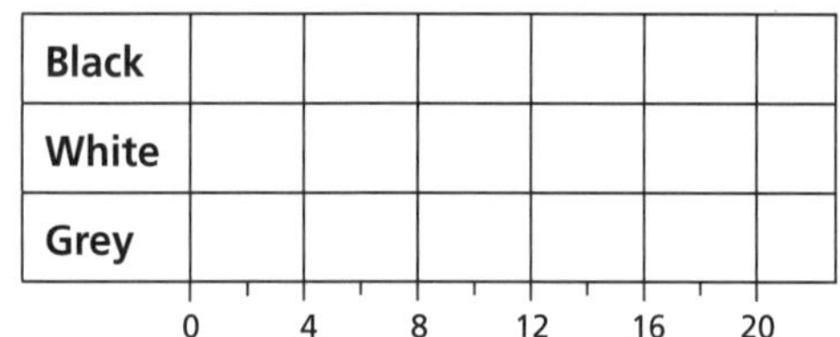

Black						
White						
Grey						
	0	4	8	12	16	20

6 Complete the pattern:

14 hundredths, 16 hundredths, 18 hundredths,

______, ______

7 Our 23 turkeys hatched 64 chicks. How many birds altogether? ______

8 a Complete this cube.

b How many edges? ______

c How many faces? ______

16:4 Extension ☐ out of 8

1 How many hours in 3 days? ______

2 I did 30 minutes of homework each night for 4 nights. How much homework did I do? ______

3 Am I a prism, a pyramid or neither if all my faces are triangles? ______

4 How many triangular faces are on a:

a square pyramid? ______

b triangular pyramid? ______

5 How many corners has a square prism? ______

6 This pyramid is made of 50c coins.

What is the total value if it is:

a 3 layers high? ______

b 4 layers high? ______

7 I have 2 surfaces, 1 edge and I can roll. What shape am I? ______

8 $2 \times 6 \times 3$ ______

Challenge

Draw a pyramid. Label and describe its features.

______

______

______

______

______

To find the change from $2 when 15c is spent, start at 15c and count on as you give the change. (We say: 20c, 30c, 50c, $1, $2.)

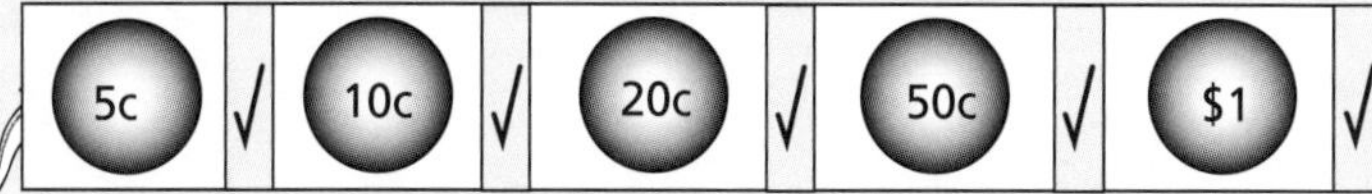

This shows you how to give change by counting on.

Find the change from $2 if you spend:

a $1.45 ______ **b** $0.90 ______

c $1.35 ______ **d** $0.75 ______

e $0.40 ______ **f** $1.05 ______

17:1 ☐ out of 15

1. 0×3 ____
2. 2×3 ____
3. 4×3 ____
4. 8×3 ____
5. $46 + 72$ ____
6. $6 + 6$ ____
7. $7 + 6$ ____
8. $15 +$ ____ $= 20$
9. $13 +$ ____ $= 20$
10. $53 + 75$ ____

11. **a** Decimal shaded. ____
    b Decimal not shaded. ____

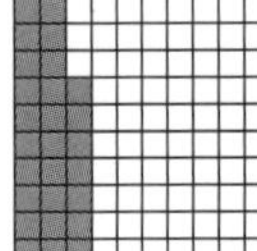

12. True or false? 0·34 is larger than 3·4. ____

13.

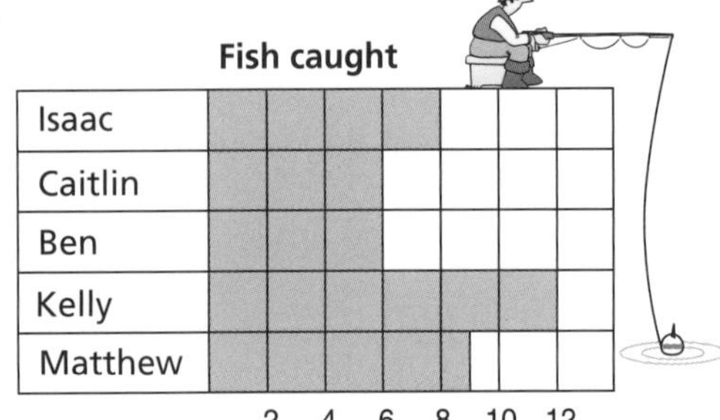

a What is the difference between the number of fish caught by Kelly and the number caught by Caitlin? ____
b Who caught 9 fish? ____

14. Jarrod got 16 runs on Monday and 23 runs on Tuesday. How many runs did he get altogether? ____

15. Draw 3 groups of 6 circles. $3 \times 6 =$ ____

17:2 ☐ out of 19

1. 5×3 ____
2. 5×6 ____
3. 7×3 ____
4. 7×6 ____
5. $71 + 54$ ____
6. 30 mL + 25 mL ____
7. 50 mL − 16 mL ____
8. $14 +$ ____ $= 30$
9. $18 +$ ____ $= 30$
10. $84 + 72$ ____

11. Write as a decimal:
    a 7 hundredths ____
    b 37 hundredths ____

12. **a** This model shows ____ hundredths.
    b Write this as a decimal. ____

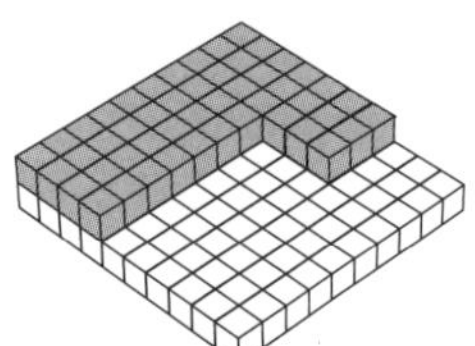

13. Suzie found 9 snails, Sarah found 15 and Scarlett found 12. How many did they find altogether? ____

14. **a** 3, 6, 9 ____, ____, ____, ____
    b 6, 12, 18 ____, ____, ____, ____

15. Is × 6 double × 3? ____

16. I had $83 and spent $35. What is my change? ____

17. How many millilitres in:
    a 3 L? ____ **b** 6 L? ____
    c 7·5 L? ____ **d** 9·25 L? ____

18. Write the short form for:
    a 6 litres ____ **b** 83 millilitres ____

19. **a** 3×6 ____ **b** 6×6 ____

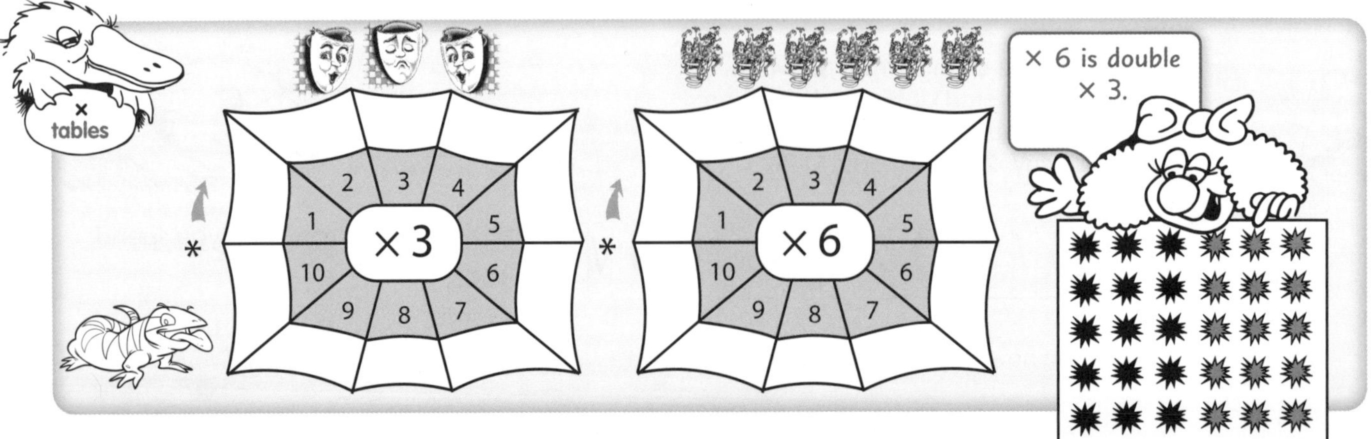

 • *AUSTRALIAN SIGNPOST MATHS 4 MENTALS* • ISBN 978 0 6557 0884 1

Answers

ID Card Answers

ID Card A

1 metres **2** centimetres **3** millimetres **4** square metres
5 square centimetres **6** litres **7** millilitres **8** kilograms **9** grams
10 hours **11** minutes **12** seconds **13** degrees Celsius
14 freezing point of water **15** boiling point of water **16** before noon
17 after noon **18** digital watch **19** analog clock **20** 5×7 **21** $20 \div 5$
22 $18 \div 6$ **23** counting numbers **24** even numbers **25** odd numbers
26 ordinal numbers **27** digits **28** picture graph **29** tally
30 column graph

ID Card B

1 horizontal line **2** vertical line **3** parallel lines **4** perpendicular lines
5 number line **6** axis of symmetry **7** tessellation **8** flip or reflect
9 translation **10** turn or rotation **11** acute angle **12** right angle
13 straight angle **14** obtuse angle **15** reflex angle **16** revolution
17 vertex **18** arm **19** number expander **20** abacus **21** 5 **22** 24
23 3 **24** 0·21 **25** 21 **26** decimal point **27** calendar **28** compass
29 coordinates **30** metre

ID Card C

1 oval **2** triangle **3** square **4** rectangle **5** rhombus **6** trapezium
7 parallelogram **8** quadrilaterals **9** pentagons **10** hexagons **11** octagon
12 kite **13** regular shapes **14** irregular shapes **15** diagonals **16** face
17 corner (or vertex) **18** edge **19** base **20** flat surface **21** curved surface
22 sphere **23** cube **24** cylinder **25** cone **26** pyramid **27** prism
28 net of a cube **29** net of a square pyramid **30** net of a triangular prism

1:1

1 5 **2** 14 **3** 47 **4** 15 **5** 89 **6** 22
7 11 **8** 45 **9** 50 **10** 84 **11** 42 926
12 $\frac{4}{10}, \frac{5}{10}, \frac{6}{10}, \frac{7}{10}, \frac{8}{10}$
13 46 716 will be circled. **14** 143
15 10c and 5c coin will be coloured. **16** unlikely
17

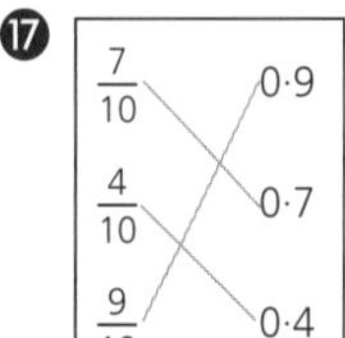

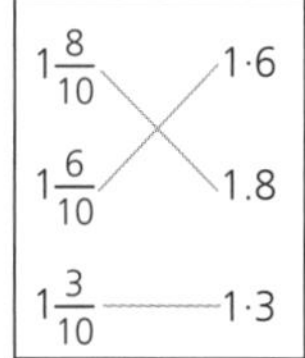

1:2

1 33 **2** 20 **3** 17 **4** 39 **5** 79 **6** 7 **7** 15 **8** 639 **9** 516
10 76 **11** 0·7
12 65

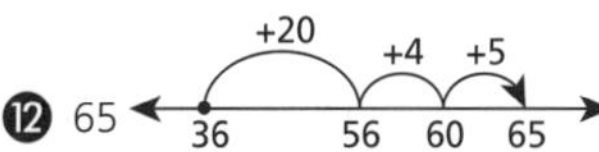

13 0·9 **14** a December 12 b December 28
15

	Thousands	Hundreds	Tens	Ones
a	846	3	9	3
b	80	4	7	5
c	87	6	3	7

Activity

10, 2, 6, 9, 4, 8, 12, 5, 11, 7
8, 11, 3, 6, 12, 5, 10, 13, 7, 9
9, 12, 4, 7, 13, 6, 11, 14, 8, 10

1:3

1 a 0·8 b 0·2 c 0·7 d 8·6 e 3·5
2 3 parts of the rectangle will be coloured.
3 a $1\frac{8}{10}$ b $3\frac{9}{10}$ c $6\frac{7}{10}$
4 50c, 10c, and 5c coin will be coloured. **5** 20
6 a 65 b 43 c 78 d 43 **7** square pyramid **8** 6

1:4

1 a 4 b 10 c 20 d 40 **2** no **3** 325 **4** 64 **5** 44
6 a 3 b 2

Challenge

Answers will vary. For example, 426 930 has 4 hundred thousands, 426 thousands, 42 693 tens. It is an even number, it is higher than 426 922 and lower than 426 936. It rounds to 427 000 (to the nearest thousand). The number before it is 426 929. It is in the pattern 426 920, 426 925, 426 930. 426 000 + 930 = 426 930

Activity

Answers will vary.

2:1

1 45 **2** 20 **3** 77 **4** 34 **5** 85 **6** 92 **7** 80 **8** 6 **9** 18
10 33 **11** a cube, 6 faces b square pyramid, 5 faces
12 a $\frac{3}{10}$ b $\frac{6}{10}$ c $\frac{9}{10}$
13 $20 **14** a 83, 93, 103, 113, 123 b 50, 55, 60, 65, 70
c 560, 570, 580, 590 d 22, 20, 18, 16, 14 **15** yes **16** 32 242
17 15 m **18** 56 722

2:2

1 78 **2** 62 **3** 79 **4** 48 **5** 68 **6** 7 **7** 12 **8** 4 **9** 7
10 87 **11** a 27 minutes past 9 b nine twenty-seven
12 a 2:22, 22 past 2 b 7:49, 11 to 8 **13** This is a triangular pyramid. It has 4 faces, 6 edges and 4 corners. The cross-section is a triangle.
14 no **15** Thursday **16** September **17** 50

Activity

even − even = even
odd − even = odd
0, 3, 5, 8, 4, 6, 2, 7, 3, 9
7, 3, 10, 2, 6, 8, 4, 9, 5, 1

2:3

1 5000 **2** 16 **3** 67 417 **4** 2 rows of 7, Area = 14 cm^2
5 a 20 to 4 b 3:40
6 a 5 minutes to 4 or 3:55 b 7 minutes past 7 or 7:07 **7** Add 20.
8 6492 **9** The 3rd triangle will be crossed out. **10** a 60 g b 64 kg

2:4

❶ 25 ❷ 94 ❸ 120 ❹ 160 ❺ 15 ❻ **a** 6:10 **b** 9:27
❼ 156 weeks ❽ 405 ❾ Tuesday

Challenge

Answers will vary.

a Questions will equal 22.

b Questions will equal 38.

c Questions will equal 40.

Activity

9, 11, 15, 12, 8, 7, 13, 10, 5, 14
14, 16, 20, 17, 13, 12, 18, 15, 10, 19
10, 13, 8, 15, 11, 6, 14, 9, 12, 16

3:1

❶ 45 ❷ 12 ❸ 24 ❹ 30 ❺ 78 ❻ 5 ❼ 41 ❽ 40
❾ 20 ❿ 48 ⓫ 3 squares out of 8 will be shaded. ⓬ 50 ⓭ 7
⓮ Answers will vary. ⓯ 43 ⓰ 4 ⓱ 2418 ⓲ 15748 ⓳ 5:59

3:2

❶ 63 ❷ 70 ❸ 30 ❹ 18 ❺ 53 ❻ 5 ❼ $3.60 ❽ 35
❾ 60 cm ❿ 129 ⓫ 9374 ⓬ 400 mL ⓭ 56946
⓮ 7 rectangles out of 10 will be shaded. ⓯ 41534
⓰ 7000 ⓱ **a** A, B and D **b** A and C **c** C **d** C ⓲ 8000 mL

Activity

(5) rhombus (6) trapezium (7) parallelogram (8) quadrilaterals

(9) pentagons (10) hexagons (11) octagon (12) kite

(13) regular shapes (14) irregular shapes

3:3

❶ **a**  **b** 

❷ **a** 9 L **b** 23 kg **c** 13 m **d** 14 mm **e** 13 cm^2
❸ 3 ❹ 6 moon shapes will be circled. $\frac{6}{12}$ or $\frac{1}{2}$ ❺ 47306
❻ **a** 637818 **b** 630602 ❼ 30 ❽ $4.40

3:4

❶ 17 ❷ **a** 8 of the 16 squares will be coloured red.
b 4 of the 16 squares will be coloured blue. **c** $\frac{4}{8}$ or $\frac{1}{2}$ ❸ **a** 9 **b** 19
❹ 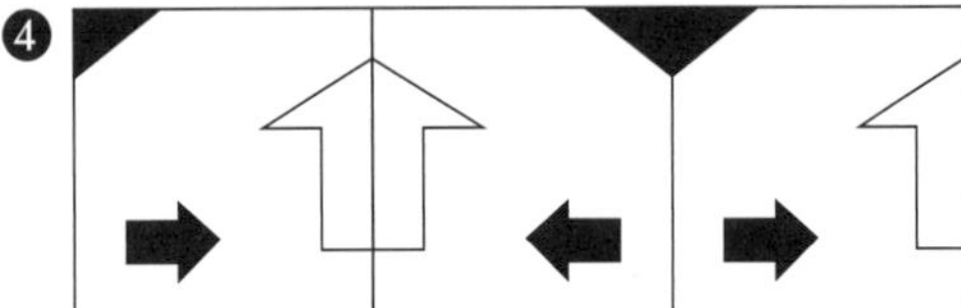

Challenge

Answers will vary.

Activity

13, 8, 16, 12, 14, 9, 17, 15, 11, 18
18, 13, 11, 9, 17, 12, 14, 19, 15, 16
12, 16, 10, 13, 7, 11, 15, 17, 9, 14

4:1

❶ 59 ❷ 36 ❸ 60 ❹ 80 ❺ 51 ❻ 4 ❼ 2 ❽ 5 ❾ 40
❿ 32 ⓫ 98411 ⓬ 60 ⓭ 85382 ⓮ 600 ⓯ **a** 12, 15, 18, 21, 24
b 16, 20, 24, 28, 32 **c** 36, 45, 54, 63, 72 ⓰ 4246, 4265, 4286
⓱ 30c ⓲ 3

4:2

❶ 35 ❷ 15 ❸ 30 ❹ 10 ❺ 40 ❻ 6 ❼ 18 ❽ 35 ❾ 30
❿ 21 ⓫ 8194 ⓬ **a** 42, 46, 50, 54 **b** 24, 26, 28, 30
⓭ 59465, 57290, 62746, 60192, and 56003 will be circled.
⓮ 15c ⓯ 6 ⓰ 500 ⓱ **a** $\frac{5}{16}$ **b** $\frac{11}{16}$

Activity

0, 6, 10, 14, 20, 16, 4, 12, 8, 18
0, 12, 20, 28, 40, 32, 8, 24, 16, 36

4:3

❶ 53628 ❷ 76483, 79802, 76394, 75000, and 84476 will be circled. ❸ **a** 24, 30, 36, 42, 48 **b** 28, 35, 42, 49, 56
c 50, 55, 60, 65 ❹ 24 ❺ 2 groups of 2 = 4, 3 groups of 2 = 6
❻ 4 ❼ 4567 ❽ five thousand, three-hundred and nine
❾ 8500 ❿ 3588
⓫ 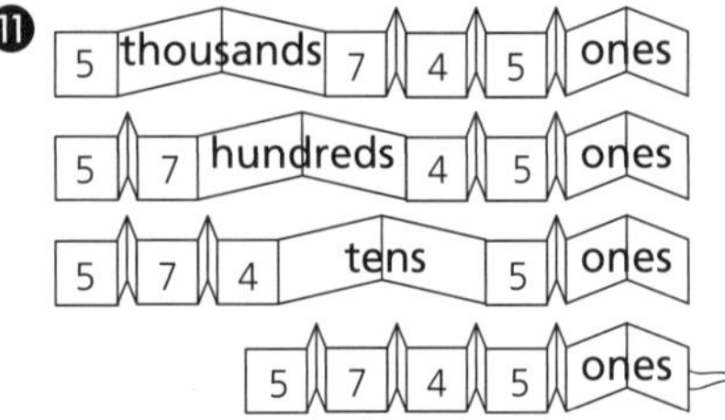

4:4

❶ **a** 52, 65, 78, 91 **b** 60, 75, 90, 105
❷ **a** $2.80 **b** $3.50 ❸ 68 ❹ **a** 9:15 **b** 1:15
❺ $2\frac{1}{2}$ ❻ 7 ❼ 39, 50, 61, 72, 83 ❽ 45 ❾ $75, $12 ❿ 10

Challenge

Answers will vary, e.g.
Add 12: 12, 24, 36, 48, 60
Subtract 10: 113, 103, 93, 83

Activity

10, 20, 30, 40, 50, 60, 70, 80, 90, 100
5, 10, 15, 20, 25, 30, 35, 40, 45, 50

5:1

❶ 60 ❷ 40 ❸ 84 ❹ 63 ❺ 98 ❻ 20, 25 ❼ 12, 15
❽ 16, 20 ❾ 53, 63 ❿ 97 ⓫ 50000 ⓬ $\frac{2}{6}$
⓭ **a** 40, 50, 60, 70 **b** 8, 10, 12, 14, 16 ⓮ $6
⓯ **a** 45 minutes **b** 50 minutes ⓰ 4 ⓱ 16

5:2

1 40 **2** 16 **3** 70 **4** 25 **5** 56 **6** 30 **7** 70 **8** 35

9 24 **10** 95 **11** 45 830, 48 273, 51 394, and 53 674 will be circled.

12 **a** $\frac{6}{8}$ **b** $\frac{7}{8}$

13 **a** 108, 117, 126, 135 **b** 72, 78, 84, 90

14 **a** 4 squares out of 6 will be coloured.
b 2 squares out of 6 will be coloured. **c** $\frac{4}{6}$

15 3 past 9 **16** 24

Activity

0, 30, 50, 70, 100, 80, 20, 60, 40, 90

0, 15, 25, 35, 50, 40, 10, 30, 20, 45

5:3

1 10, 15 **2** 71 254, 65 488, 72 465, and 70 896 will be circled.

3 **a** $\frac{5}{10}$ or $\frac{1}{2}$ **b** $\frac{5}{10}$ or $\frac{1}{2}$

4 5:11 **5** **a** 12:41 **b** 19 to 1

6 **a**  **b** 4:25 **7** **a** 60 **b** 24 **c** 7

8 7 of the 8 sections will be coloured blue.

5:4

1 9 of 12 parts will be shaded. **2** **a** 4:03 or 3 past 4 **b** 3:58 or 2 to 4

3 $5.75 **4** 55 **5** 21 **6** 20 **7** 6019 **8** 40

Challenge

Answers will vary, e.g., $\frac{14}{15}$

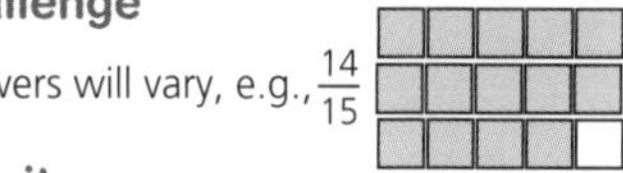

Activity

	2	4	6	8	10
× 2	4	8	12	16	20

	2	4	6	8	10
× 5	10	20	30	40	50

	1	2	3	4	5
× 4	4	8	12	16	20

	1	2	3	4	5
× 10	10	20	30	40	50

	6	3	8	7	9
× 0	0	0	0	0	0

	4	7	3	9	6
× 1	4	7	3	9	6

6:1

1 4 **2** 16 **3** 24 **4** 12 **5** 32 **6** 16 **7** 28 **8** 24 **9** 8

10 44 **11** $\frac{5}{8}$ **12**

13 **a** 43 mm **b** 79 mm **14** 15 **15** **a** <u>3</u> × 5 = 15 **b** <u>9</u> × 5 = 45
c <u>10</u> × 5 = 50 **d** <u>7</u> × 5 = 35

16

	7	10	2	5	1	3	8	6	9
× 5	35	50	10	25	5	15	40	30	45

6:2

1 28 **2** 40 **3** 20 **4** 32 **5** 21 **6** 36 **7** 32 **8** 40 **9** 12

10 22 **11** 8, 12 **12** $\frac{2}{9}$ **13** 2:38

14 **a** 3:15 **b** 3:30 **15** 40 mm

16 **a** 4 **b** 3 **17** 12

Activity

13, 8, 16, 12, 14, 9, 17, 15, 11, 18

18, 13, 11, 9, 17, 12, 14, 19, 15, 16

12, 16, 10, 13, 7, 11, 15, 17, 9, 14

6:3

1 3 frogs will be green, and 5 frogs will be red.

2 **a** **b** **3** 25 minutes **4** 6 cm

5 **a** 8 cm 4 mm **b** 5 cm 9 mm **6** 36 209 **7** **a** 2, 2 **b** 8, 8 **8** 25

9

	7	10	2	5	1	3	8	6	9
× 10	70	100	20	50	10	30	80	60	90

10 16,

6:4

1 **a** 899 **b** 696 **c** 193 **2** 26 **3** 72 minutes or 1 hour 12 minutes

4 34, 51, 68 **5** 6 of 8 triangles will be coloured.

6 **a** 2 m 99 cm or 299 cm **b** 2 m 97 cm or 297 cm
c 3 m 87cm or 387 cm **d** 4 m 45cm or 445 cm

Challenge

Answers will vary.

Activity

12, 4, 24, 20, 32, 8, 36, 28, 16, 40

7:1

1 30 **2** 28 **3** 90 **4** 15 **5** 22 **6** 583 **7** 485 **8** 20

9 24 **10** 33 **11** 325 cm

12 15,

13 **a** 6 **b** $1\frac{2}{4}$ or $1\frac{1}{2}$

14 **a** 6 cm **b** 1 cm **15** **a** <u>3</u> groups of <u>6</u> **b** <u>3</u> × <u>6</u> = <u>18</u>

7:2

1 36 **2** 12 **3** 28 **4** 32 **5** 88 **6** 50 **7** 4 **8** 35 **9** 30

10 39 **11** 27 mm **12** Estimates will vary, 6 cm, 60 mm

13 $\frac{5}{8}$, $1\frac{1}{4}$

⓮ a 9 cm 3 mm b 7 cm 9 mm ⓯ 4 cm^2

⓰ 5 and a half balls will be coloured.

Activity

a <u>8</u> ÷ <u>2</u> b <u>27</u> ÷ <u>3</u> c 4 d 9 e 6 f 6 g 9 h 7

7:3

❶ a 4 m 59 cm b 3 m 27 cm ❷ 28 cm

❸ $\frac{7}{3}$, $2\frac{1}{3}$ ❹ a $3\frac{1}{2}$ b $1\frac{3}{4}$ ❺ 6 cm^2 ❻ square centimetres

❼ a 95 mm b 42 mm ❽ a $2\frac{2}{5}$ b $2\frac{2}{3}$

❾ 20 219

❿

	7	10	2	5	1	3	8	6	9
× 4	28	40	8	20	4	12	32	24	36

7:4

❶ a 9 cm 9 mm b 9 mm ❷ 36 ❸ a 10 past 7 or 7:10
b 7:12 or 12 past 7 ❹ B ❺ a 128 b 96 c 160

Challenge

Answers will vary. Fractions may include $1\frac{1}{3}$, $1\frac{3}{5}$, etc.

Activity

a 3 rows of <u>3</u> = <u>9</u> cm^2

b <u>3</u> rows of <u>2</u> = <u>6</u> cm^2

c <u>2</u> rows of <u>5</u> = <u>10</u> cm^2

8:1

❶ 53 ❷ 67 ❸ 69 ❹ 86 ❺ 77 ❻ 5 ❼ 8 ❽ 47 ❾ 36

❿ 96 ⓫ $\frac{12}{5}$, $2\frac{2}{5}$ ⓬ Estimates will vary, 5 cm, 50 mm

⓭ $\frac{4}{10}$, $\frac{5}{10}$, $\frac{6}{10}$, $\frac{7}{10}$, $\frac{8}{10}$

⓮ 4 and a half ice creams will be coloured. ⓯ a 20 b 28 ⓰ 11

8:2

❶ 77 ❷ 49 ❸ 59 ❹ 96 ❺ 25 ❻ 5 ❼ 15 ❽ 31 ❾ 69

❿ 55 ⓫ 17 ⓬ $\frac{5}{4}$, $1\frac{1}{4}$ ⓭ 12 cm^2 ⓮ a $2\frac{1}{3}$ b $2\frac{2}{4}$ or $2\frac{1}{2}$ ⓯ 33

⓰ 88

Activity

a 7 cm^2 b 5 cm^2 c 11 cm^2

8:3

❶ 77 ❷ 31 ❸ $\frac{9}{5}$, $1\frac{4}{5}$ ❹ Answers may vary. (true)

❺ a B b C ❻ a B b C ❼ a b a triangle

8:4

❶ 6 ❷

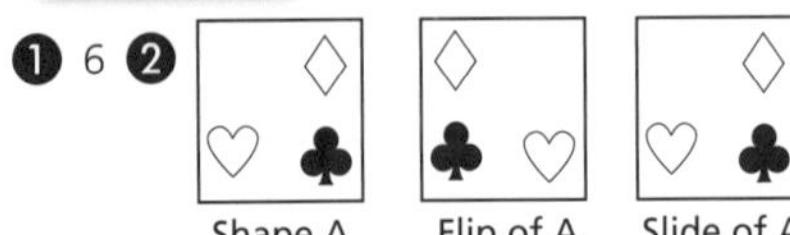

Shape A Flip of A Slide of A

❸ hexagon ❹ circle ❺ squares and rectangles ❻ 80

Challenge

4-sided shapes will be drawn. Students may draw regular and/or irregular quadrilaterals. See ID card C (3) – (8) and (12) for examples.

Activity

a 17 b 31 c 15 d 47 e 41 f 71 g 65 h 27 i 33 j 51

9:1

❶ 10 ❷ 20 ❸ 16 ❹ 32 ❺ 88 ❻ 4 ❼ $15 ❽ 8 ❾ 14

❿ 57 ⓫ a $2\frac{5}{6}$ b $5\frac{1}{4}$ ⓬ $\frac{4}{6}$, $\frac{5}{6}$, $\frac{6}{6}$, $\frac{7}{6}$, $\frac{8}{6}$

⓭ a C b B ⓮ right angle ⓯ 39 218 ⓰ 35 335, 35 782, 35 836

⓱ 3 lines of symmetry will be drawn on the circle.

9:2

❶ 89 ❷ 30 ❸ 30 ❹ 15 ❺ 83 ❻ 45 ❼ 90 ❽ 50 ❾ 28

❿ 91 ⓫ 1 ⓬ C ⓭ flip

⓮

Smiley face	Circle	Square	Rhombus
Happy	3	5	2
Sad	5	1	2

⓯ 34 029 ⓰

Activity

a 46 673, 46 573, 46 583, 45 583

b 368 025, 366 025, 366 425, 366 405, 367 405

c 209 871, 209 471, 207 471, 207 431, 204 431

9:3

❶ 60 ❷ 71 ❸ C, B, A

❹

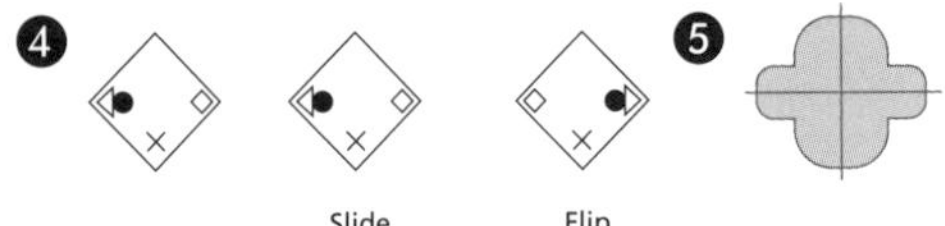

❺

❻ 81 ❼ $6\frac{1}{2}$ ❽ 79 463, 79 465, 79 640 ❾ 42 739 ❿ 4000

9:4

❶

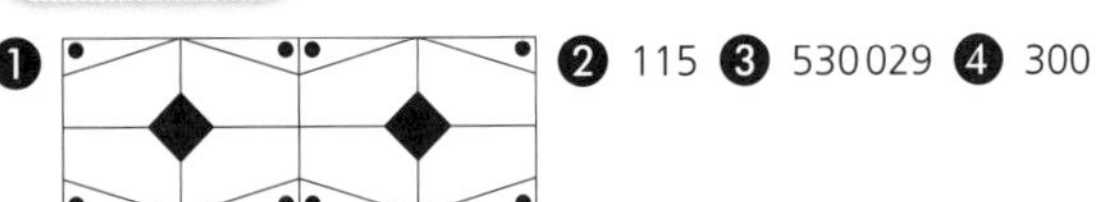

❷ 115 ❸ 530 029 ❹ 300 000

❺ 25 ❻ 6 ❼ a 3 cm b 9 cm

Challenge

Answers will vary, e.g.

It is a 3-digit number. The number after it is 839. When it is rounded to the nearest hundred, it is rounded down to 800. 800 + 38 = 838, 900 − 62 = 838.

Activity

a 28 468, 28 368, 28 378, 28 377

b 298 457, 294 457, 294 757, 294 717, 294 917

c 956 292, 956 252, 956 255, 956 225, 958 225

10:1

❶ 9 ❷ 5 ❸ 9 ❹ 17 ❺ 8 ❻ 12 ❼ 15 ❽ 101 m ❾ 60 ❿ 75

⓫ 91 708, 92 087, 92 807 ⓬ 26 184 ⓭ 900

⓮ a 2 turtles will be coloured. b 2 turtles will be circled. c yes

⓯ $\frac{4}{7}$, $\frac{5}{7}$, $\frac{6}{7}$, $\frac{7}{7}$ ⓰ a 5 b 3 ⓱ 32 071

 ISBN 978 0 6557 0884 1

10:2

❶ 10 ❷ 20 ❸ 17 ❹ 20 ❺ 71 ❻ 79 ❼ 90c ❽ 10
❾ 6 ❿ 14
⓫

Shapes	Rhombus	Pentagon	Rectangle
Shaded	3	3	2
Not shaded	5	3	4

⓬ 28 098, 28 439, 28 463 ⓭ a true b true ⓮ 9000 ⓯ 483 403
⓰ 2 heads, 2 tails or 1 head and 1 tail ⓱ 14

Activity

Yes.

Answers will vary.

10:3

❶ 53 ❷ 62 ❸ 15 794 ❹ 5000 ❺ half turn
❻ $\frac{4}{9}, \frac{5}{9}, \frac{6}{9}, \frac{7}{9}$ ❼ 67 502, 67 590, 75 580 ❽ a 2 b 3
❾ a Answers may vary (e.g. $\frac{2}{4}, \frac{4}{8}, \frac{3}{6}$). b Answers may vary (e.g. $\frac{1}{1}, \frac{2}{2}, \frac{3}{3}$).
c Answers may vary (e.g. $\frac{8}{10}, \frac{12}{15}, \frac{16}{20}$).
d Answers may vary (e.g. $\frac{1}{2}, \frac{2}{4}$).
❿ 1009, 1010, 1011 ⓫ 56 619

10:4

❶ a 412 050 b 45 035 c 37 009
❷
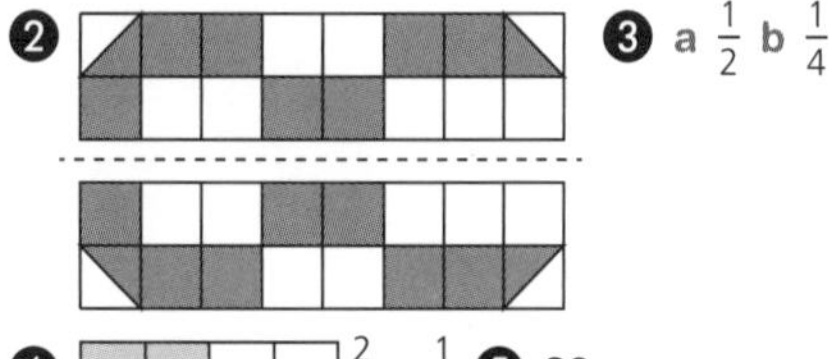
❸ a $\frac{1}{2}$ b $\frac{1}{4}$ c $\frac{3}{4}$
❹ $\frac{2}{4}$ or $\frac{1}{2}$ ❺ 38

Challenge

Answers will vary, e.g. $\frac{2}{4}, \frac{3}{6}, \frac{4}{8}, \frac{5}{10}$. The fractions will be drawn.

Activity

a 3, 1 b 4, 2 c 6, 3 d 9, 3

11:1

❶ 10 ❷ 20 ❸ 15 ❹ 30 ❺ 80 ❻ 9 ❼ 16 ❽ 17
❾ 7 ❿ 18 ⓫ a true b true ⓬ $\frac{4}{8}, \frac{5}{8}, \frac{6}{8}, \frac{7}{8}, \frac{8}{8}$
⓭ A 60°C B 30°C C 50°C ⓮ 80c ⓯ 5, 10, 15, 20, 25, 30, 35, 40
⓰ 12, ○ ○○ ○ ○○ ○ ○○ ○ ○○

11:2

❶ 14 ❷ 28 ❸ 40 ❹ 80 ❺ 82 ❻ 32 ❼ 3 ❽ 8 ❾ 4 ❿ 24
⓫ a $\frac{3}{4}$ b $\frac{2}{6}$ ⓬ 0, $\frac{1}{4}, \frac{2}{4}, \frac{3}{4}$, 1, $1\frac{1}{4}$, $1\frac{2}{4}$
⓭ A ⓮ yes ⓯ 53°C ⓰ 7, 9, 11, 13, 15
⓱ a no b yes ⓲ a 12 b 4 c 3 d 3 ⓳ 642 309

Activity

Answers will vary, e.g. flipping heads on a coin, picking a red card from a deck of playing cards, rolling a 2, 4 or 6 on a dice.

11:3

❶ 85 ❷ 74 ❸ a true b false
❹ $2\frac{1}{2}$, 3, $3\frac{1}{2}$, 4, $4\frac{1}{2}$, 5 ❺ a A b B
❻ a b c
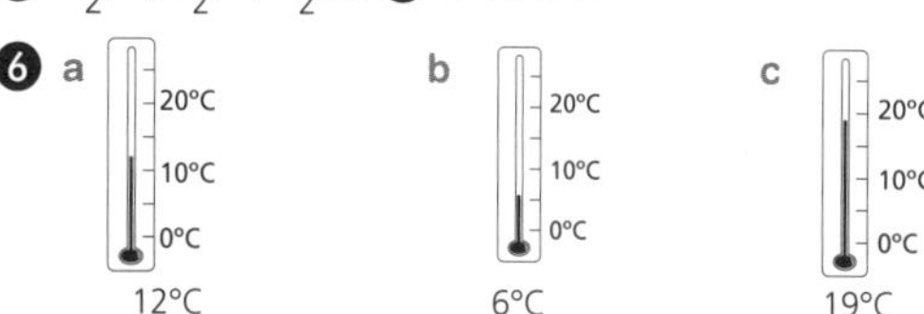

11:4

❶ 7 ❷ a $\frac{3}{10}$ b $\frac{3}{4}$ (or $\frac{75}{100}$) ❸ D ❹ a A b B ❺ 51, 53, 55, 57, 59

Challenge

a 28 + 9 = 30 + 7 b 37 + 6 = 40 + 3 c 45 + 9 = 50 + 4
d 56 + 8 = 60 + 4 e 84 + 8 = 90 + 2 f 76 + 7 = 80 + 3
g 57 + 6 = 60 + 3 h 49 + 6 = 50 + 5 i 47 + 7 = 50 + 4
j 67 + 6 = 70 + 3

Activity

10, 2, 5, 1, 6, 9, 4, 8, 3, 7
8, 7, 6, 9, 1, 5, 10, 3, 0, 4
6, 3, 1, 5, 9, 7, 2, 10, 4, 8

12:1

❶ 25 ❷ 15 ❸ 5 ❹ 19 ❺ 75 ❻ 6 ❼ 21 ❽ 16 ❾ 6
❿ 26 ⓫ 17°C ⓬ 43 564 ⓭ 82 (+30, +3, +2; 47, 77, 80, 82)
⓮ 231 508 ⓯ a 5000 b 4670 ⓰
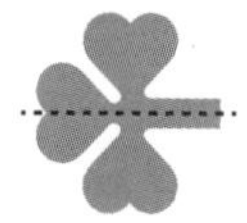
⓱ 2 out of 4 ⓲ shape A

12:2

❶ 32 ❷ 32 ❸ 55 ❹ 60 ❺ 84 ❻ 37 ❼ 10 ❽ 18
❾ 25 ❿ 15 ⓫ Yes ⓬ a 3 b 6 c 4 d 5
⓭ four thousand and ninety-three
⓮ a 8 000 000 b 7 835 000 ⓯ 7 689 023, 7 655 982, 7 645 890
⓰ a 65
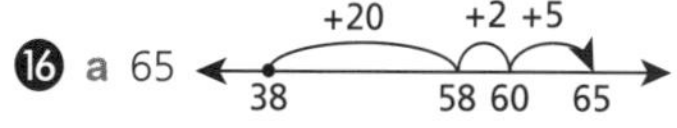

b 25
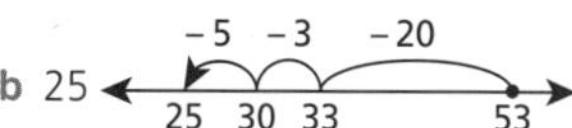

Activity

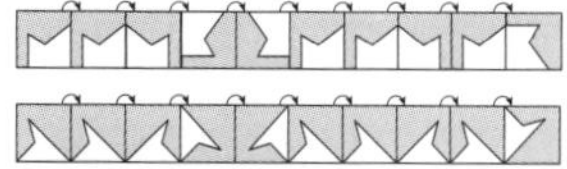

12:3

❶ 112 ❷ 102 ❸ a 5 b 2 c 2 d 5 ❹ 325 492 ❺ 3 269 671, 3 268 980, 3 264 879 ❻ 70 000 ❼ a 6 000 000 b 5 644 000

8 hundred square, 10

9 a 86 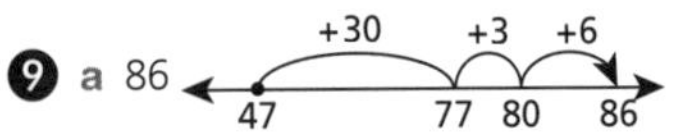

b 13 (−7 −1 −20; 13 20 21 41)

12:4

1 16, 32, 64, 128 2 141 (+40 +3 +1; 97 137 140 141)

3 5 × 5 4 5 5 2 6 a 26 b 28

7 a 7:45 b 6:45 8 3 9 986 350 10 12

Challenge

Answers will vary.

Activity

a 57 748, 56 748, 56 749, 56 739

b 365 607, 368 607, 368 407, 368 447, 368 443

c 473 996, 473 976, 473 979, 473 959, 476 959

13:1

1 4 2 8 3 8 4 16 5 17 6 80 7 0 8 24 9 24

10 81 11 578 936, 578 354, 578 099 12 a 37 500 b 38 000

13 a 35 out of 100 b 65 out of 100 14 a 16, 20, 24, 28, 32

b 12, 15, 18, 21, 24 c 24, 30, 36, 42, 48 d 32, 40, 48, 56, 64

15 16 16 15 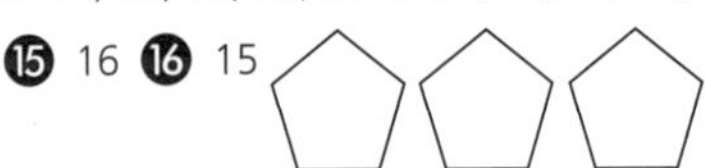

13:2

1 16 2 32 3 20 4 40 5 45 6 48 7 48 8 36 9 32

10 83 11 a 5 000 000 b 4 627 000

12 a 93 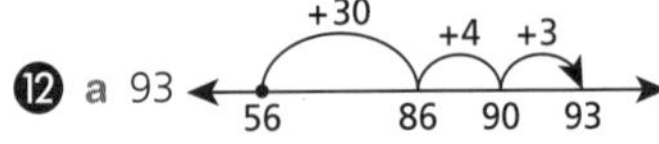

b 47 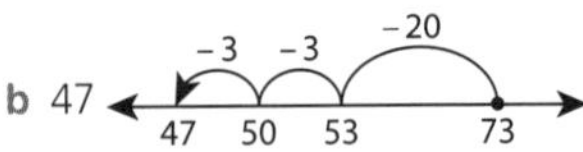

13 8 356 978, 8 356 809, 8 356 263 14 5000

15 a 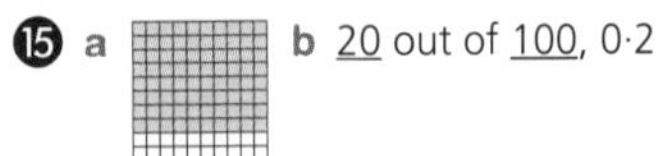 b <u>20</u> out of <u>100</u>, 0·2

Activity

×	5	10	4	0	2	6	1	3
2	10	20	8	0	4	12	2	6
4	20	40	16	0	8	24	4	12
8	40	80	32	0	16	48	8	24
5	25	50	20	0	10	30	5	15
10	50	100	40	0	20	60	10	30

13:3

1 33 2 16 3 a 284 500 b 284 000

4 a 92 (+20 +2 +2; 68 88 90 92)

b 14 (−6 −2 −40; 14 20 22 62)

5 9, 0·09 6 a 16, 20, 24, 28 b 32, 40, 48, 56

7 a 40 b 80 c 48 8 641 858 9 294 673, 294 658, 294 599

13:4

1 131 (+30 +1 +1; 99 129 130 131)

2 a 60 + <u>40</u> = 100 b 100 − 40 = <u>60</u> 3 a 28 b 41 4 38

5 1 m 50 cm or 150 cm 6 a 88, 84, 80, 76 b 76, 68, 60, 52

7 66 8 51, 53, 55, 57, 59

Challenge

a 25 + 3 = 20 + <u>8</u> b 54 + 5 = 50 + <u>9</u>

c 37 + 9 = 40 + <u>6</u> d 57 + 8 = 60 + <u>5</u>

e 46 + 6 = <u>50 + 2</u>, Answers may vary.

f 79 + 5 = <u>80 + 4</u>, Answers may vary.

g 88 + 8 = <u>90 + 6</u>, Answers may vary.

h 76 + 12 = <u>80 + 8</u>, Answers may vary.

Activity

a 23 (−7 −1 −30; 23 30 31 61)

b 29 (−1 −6 −20; 29 30 36 56)

c 36 (−4 −3 −30; 36 40 43 73)

d 27 (−3 −2 −50; 27 30 32 82)

14:1

1 22 2 11 3 16 4 32 5 96 6 8 7 10 8 40 9 80

10 16 11 <u>36</u> out of <u>100</u>, 0·36 12 a 0·7, 0·8, 0·9, 1·0

b 0·38, 0·39, 0·40 or 0·4 13 a B b C 14 a 0·5 b 0·3 15 39

16 8 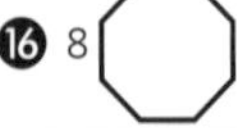

14:2

1 55 2 32 3 64 4 56 5 92 6 72 7 72 8 4 9 6

10 36 11 a 0·9 b 4·6 12 $\frac{34}{100}$, (34 out of 100), 0·34 13 0·42

14 $\frac{6}{10}, \frac{7}{10}, \frac{8}{10}, \frac{9}{10}$ 15 yes 16 cube and cylinder

17 15 minutes 18 200 19 400 20 33

 ISBN 978 0 6557 0884 1

Activity

(22) sphere (23) cube (24) cylinder (25) cone (26) pyramid (27) prism (28) <u>net</u> of a cube (29) net of a <u>square pyramid</u> (30) net of a <u>triangular prism</u>

14:3

1 39 **2** $18 **3** **a** 0·35 **b** 0·83

4 **a** $\frac{6}{10}, \frac{7}{10}, \frac{8}{10}, \frac{9}{10}, \frac{10}{10}$ or 1, $\frac{11}{10}$ **b** 0·78, 0·79, 0·80 or 0·8

5 sphere, 1 **6** **a** B **b** C **7** 78 **8** 6012 **9** 95c **10** $1.20 **11** 53

14:4

1 34 hundredths, 0·34 **2** 96 **3** **a** 4 **b** <u>4</u> + <u>4</u> – <u>6</u> = <u>2</u> **4** 48

5 **a** 71, 62, 53, 44 **b** 63, 55, 47, 39

Challenge

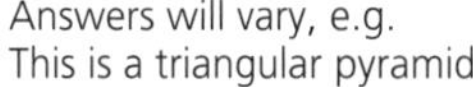

Answers will vary, e.g.
This is a triangular pyramid.
It has 4 faces, 6 edges, and 4 vertices. The cross-section of this shape is a triangle.

Activity

	2	4	6	8	10
× 4	8	16	24	32	40

	1	2	3	4	5
× 8	8	16	24	32	40

8, 16, 24, 32, 40, 48, 56, 64, 72, 80

15:1

1 24 **2** 80 **3** 28 **4** 15 **5** 93 **6** 26 **7** 40 **8** 20 **9** 12

10 18 **11** **a** 0·66 **b** 0·79 **12** **a** 0·4, 0·5, 0·6, 0·7 **b** 0·31, 0·32, 0·33

13 **a** false **b** true **14** **a** 6 **b** 8 **15** $2.15 **16** 698

15:2

1 44 **2** 10 **3** 16 **4** 32 **5** 93 **6** 46 **7** 23 **8** 48 **9** 72

10 23 **11** 0·53

12

Ones	Tenths	Hundredths
0 •	4	6

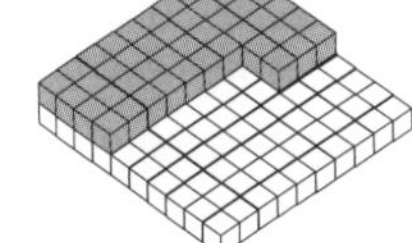

13 0·5 or 5 tenths **14** **a** C **b** C **c** A, B and D

15 **a** 449·68 **b** 727·94 **16** 36

Activity

a 11, 6, 5 **b** 15, 6, 9 **c** 24, 16, 8
d 32, 14, 18 **e** 33, 28, 5 **f** 52, 37, 15

15:3

1 9 **2** $18 **3** <u>64</u> out of <u>100</u> **4** **a** 0·37 **b** 0·8 **5** 0·6

6 B **7** 9 **8** 2 **9** 5 **10** 60

15:4

1 52 **2** **a** 83 – 22 = 81 – <u>20</u> = <u>61</u> **b** 68 – 34 = 64 – <u>30</u> = <u>34</u>

3 **a** 30 **b** 40 **c** 60 **4** **a** 4 **b** 6 **5** 84 **6** Thursday **7** 30

Challenge

Answers will vary, e.g.
This is a cube. It has
6 faces, 12 edges, and 8 vertices. The cross-section of this shape is a square.

Activity

8, 1, 5, 4, 9, 6, 3, 10, 2, 7
9, 10, 4, 7, 0, 5, 2, 8, 6, 3

16:1

1 45 **2** 48 **3** 35 **4** 32 **5** 115 **6** $12 **7** 17 **8** 5

9 43, 45 **10** 136 **11** 0·78 **12** 0·23 **13** 0·48 **14** **a** false **b** true

15 cylinder and cone **16** **a** 6 **b** 10 **c** 20

16:2

1 95 **2** 61 **3** 22 **4** 88 **5** 127 **6** 55 cm **7** $4.60

8 40 kg **9** 32 **10** 158

11  35 out of 100, (0·35 or $\frac{35}{100}$)

12 0·32 **13** 449·68 **14** 3 **15** 34 **16** 70

17 $100 + $100 + $20 + $10 +$1 + 20c + 5c

Activity

	2	4	6	8	10
× 3	6	12	18	24	30

	7	5	9	3	1
× 3	21	15	27	9	3

	1	2	3	4	5
× 6	6	12	18	24	30

	8	10	7	9	6
× 6	48	60	42	54	36

16:3

1 $96 **2** $72 **3** **a** 27 **b** 0·27 **4** prism

5

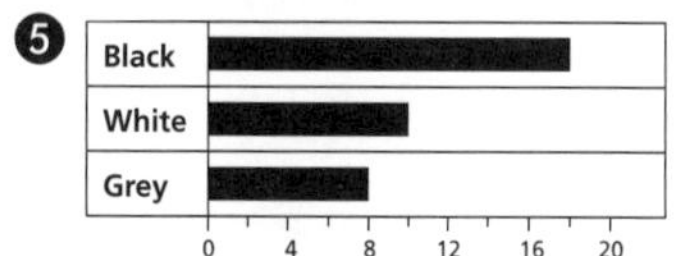

6 20 hundredths, 22 hundredths **7** 87

8 **a** 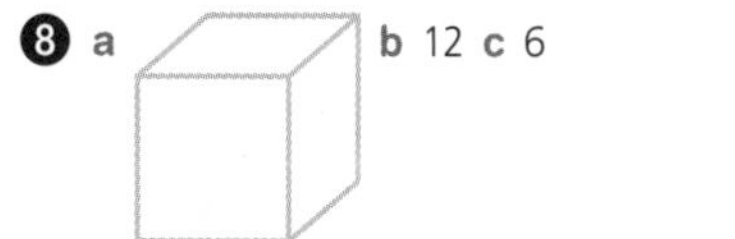 **b** 12 **c** 6

16:4

1 72 **2** 2 hours **3** pyramid **4** **a** 4 **b** 4 **5** 8

6 **a** $3 **b** $5 **7** cone **8** 36

Challenge

Answers will vary, e.g.
This is a square pyramid.
It has 5 faces, 8 edges,
and 5 vertices. It is made
up of a square base and 4 triangular sides.

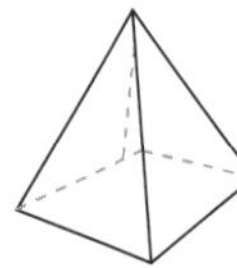

Activity

a 55c **b** $1.10 **c** 65c **d** $1.25 **e** $1.60 **f** 95c

17:1

❶ 0 ❷ 6 ❸ 12 ❹ 24 ❺ 118 ❻ 12 ❼ 13 ❽ 5 ❾ 7

❿ 128 ⓫ **a** 0·27 **b** 0·73 ⓬ false ⓭ **a** 6 **b** Matthew ⓮ 39

⓯ 18 ○○○ ○○○ ○○○ / ○○○ ○○○ ○○○

17:2

❶ 15 ❷ 30 ❸ 21 ❹ 42 ❺ 125 ❻ 55 mL ❼ 34 mL

❽ 16 ❾ 12 ❿ 156 ⓫ **a** 0·07 **b** 0·37 ⓬ **a** 49 **b** 0·49

⓭ 36 ⓮ **a** 12, 15, 18, 21 **b** 24, 30, 36, 42 ⓯ yes ⓰ $48

⓱ **a** 3000 mL **b** 6000 mL **c** 7500 mL **d** 9250 mL

⓲ **a** 6 L **b** 83 mL ⓳ **a** 18 **b** 36

Activity

3, 6, 9, 12, 15, 18, 21, 24, 27, 30
6, 12, 18, 24, 30, 36, 42, 48, 54, 60

17:3

❶ true (T) ❷ 0·76 ❸ **a** 4 L **b** 9 L ❹ **a** millilitres **b** litres

❺ **a** 12 hours **b** 61 hours ❻ **a**

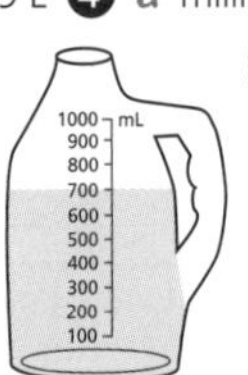

b 600 mL

❼ **a** 8 m **b** 74 mL **c** 92 L **d** 35 mm

17:4

❶ <u>36</u> and <u>37</u> ❷ 42 ❸ yes ❹ 240 ❺ 242 ❻ 40

❼ $2.80 or 280c ❽ 100 ❾ 92

Challenge

Answers will vary.

Activity

Answers will vary, e.g.
The graph represents the height of each animal in metres. The columns could represent the height of each animal.

18:1

❶ 3 ❷ 6 ❸ 6 ❹ 12 ❺ 18 ❻ 74 ❼ 43 ❽ 5 ❾ 8 ❿ 26

⓫ 30 out of 100 squares will be coloured, 0·3 ⓬ 600 mL ⓭ 1 L

⓮ Answers will vary, e.g at the corners of a door. ⓯ A ⓰ 231 629

⓱ **a** 30 **b** 60

18:2

❶ 9 ❷ 18 ❸ 30 ❹ 60 ❺ 28 ❻ 27 ❼ 27 ❽ 22 ❾ 86

❿ 15 ⓫ 16 ⓬ **a** L **b** mL **c** L **d** mL ⓭ mL or millilitres

⓮ 225 mL ⓯ **a** 4 **b** 4 **c** 2 ⓰ 30 min, 15 min

Activity

4, 28, 20, 8, 32, 0, 12, 24, 36, 16
8, 56, 40, 16, 64, 0, 24, 48, 72, 32

18:3

❶ 27 ❷ 17 ❸ 185 mL ❹ **a** 500 mL **b** 80 L **c** 1·5 m

❺ **a** C, B, A **b** B ❻ **a** 1:40 or 20 to 2 **b**

c yes ❼ 16

18:4

❶ 8 L ❷ **a**

9:00 or 9:30 **b** Yes, if the minute hand is on the 3 or the 9.

❸ **a** 11 **b** 7 ❹ **a** true **b** false **c** false

Challenge

2 × 6	12
2 × 3	6
5 × 3	15
0 × 6	0
5 × 6	30
1 × 6	6

8 × 2	16
4 × 2	8
3 × 3	9
3 × 6	18
10 × 6	60
4 × 3	12

Activity

0, 15, 24, 9, 30, 18, 27, 6, 21, 12
0, 30, 48, 18, 60, 36, 54, 12, 42, 24

19:1

❶ 9 ❷ 24 ❸ 18 ❹ 5 ❺ 27 ❻ 48 ❼ 21 ❽ $21 ❾ 2

❿ 17 ⓫ A, D, B, C ⓬ **a** 27 **b** 45

⓭ [○○○ ○○○] [○○○ ○○○] [○○○ ○○○] [○○○ ○○○] ⓮ 60 ⓯ yes ⓰ 30

⓱ true ⓲ 36, 45, 54, 63, 72

19:2

❶ 36 ❷ 54 ❸ 27 ❹ 72 ❺ 19 ❻ 45 ❼ 63 ❽ 6 ❾ 36

❿ 35 ⓫ **a** 2 tenths **b** 20 hundredths ⓬ right angle (or reflex angle)

⓭ 32 ⓮ 24 ⓯ 12 ⓰ 0 or 5 ⓱ 66 ⓲ **a**

b 4

Activity

1 × 9 = 09, 0 + 9 = <u>9</u>
2 × 9 = 18, 1 + 8 = <u>9</u>
3 × 9 = 27, 2 + 7 = <u>9</u>
4 × 9 = <u>36</u>, <u>3</u> + <u>6</u> = <u>9</u>
5 × 9 = <u>45</u>, <u>4</u> + <u>5</u> = <u>9</u>
6 × 9 = <u>54</u>, <u>5</u> + <u>4</u> = <u>9</u>
7 × 9 = <u>63</u>, <u>6</u> + <u>3</u> = <u>9</u>
8 × 9 = <u>72</u>, <u>7</u> + <u>2</u> = <u>9</u>
9 × 9 = <u>81</u>, <u>8</u> + <u>1</u> = <u>9</u>
10 × 9 = <u>90</u>, <u>9</u> + <u>0</u> = <u>9</u>

The digits in each answer always add up to <u>9</u>.

19:3

1 34 **2** 17 **3** right angle **4** **a** 42 **b** 64 **5** **a**  **b** 3

6 **a** yes **b** yes **7** B will be circled.

8 **a** 15 **b** 35 **9** **a** 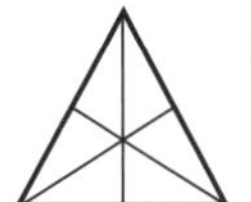 **b** no

19:4

1 regular hexagon **2** **a** 72 **b** 88 **3** **a** 12 **b** 12 **4** 5 

5 9 **6** 8

Challenge

Answers will vary.

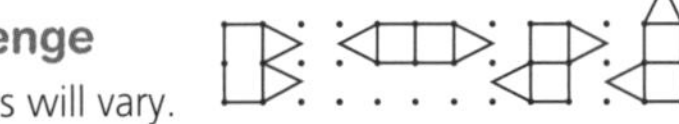

Activity

9, 18, 27, 36, 45, 54, 63, 72, 81, 90
9, 45, 72, 18, 63, 81, 27, 54, 90, 36

20:1

1 0 **2** 17 **3** 90 **4** 27 **5** 451 **6** 4 **7** 5 **8** 70

9 4 **10** 933 **11** **a** 20 **b** 28 **12** 4 **13** 0 **14** 48

15 **a** December 18 **b** December 13 **16** 6300 **17** 56 450 **18** $0.85

20:2

1 24 **2** 48 **3** 4 **4** 4 **5** 957 **6** 0·5, 0·6 **7** 48 **8** 27

9 3 **10** 822 **11** **a** 18 **b** 48 **12** $1008 **13** **a** Town Hall
b Fire Station **c** 1 cm = 100 m **d** 200 m

14 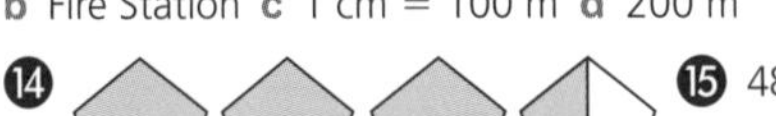 **15** 48

Activity

(16) face (17) corner (or vertex) (18) edge (19) base (20) flat surface
(21) curved surface (22) sphere (23) cube

20:3

1 401 **2** $68 **3** **a** 8 **b** 24 **4** $1.75 **5** **a** 5 **b** 10 **c** 15 **d** 50

6 10 each **7** 57 492 **8** 698 **9** 234 597 **10** **a** 6 **b** 12

11 three hundred and forty-nine **12** 854 **13** 1·9, 2·0, 2·1, 2·2

20:4

1 **a** 12 **b** 24 **2** 4 **3** 2 **4** **a** 128 **b** 96

5 13 **6** 4 **7** **a** 2 **b** 3

Challenge

3 × 2	6
5 × 4	20
2 × 10	20
2 × 5	10
3 × 8	24
5 × 8	40

6 × 10	60
5 × 5	25
8 × 5	40
4 × 8	32
1 × 10	10
9 × 8	72

Activity

5

21:1

1 27 **2** 24 **3** 16 **4** 50 **5** 834 **6** 45 **7** 18 **8** 3 **9** 30

10 913 **11** 92 **12** impossible **13** blue **14** 68 371

15 **a** 400 **b** 900 **16** 64 729 **17**

18 **a** July **b** April **19** 125 cents **20** 35 671

21:2

1 40 **2** 8 **3** 42 **4** 6 **5** 979 **6** 18 **7** 3 **8** 7 **9** 7

10 841 **11** 730 **12** **a** 2 km **b** 7 km **c** Red St **d** the shop

13 no (even chance) **14** Wednesday **15** 46·67 **16** 4:50

Activity

7, 3, 10, 5, 6, 9, 1, 8, 2, 4
7, 2, 6, 3, 10, 4, 9, 5, 8, 0

21:3

1 200 **2** 568 **3** $317 **4** **a** 50 **b** 75 **5** impossible

6 **a** C **b** A **7** 42 **8** December **9** **a** cm **b** kg **c** L **d** min

21:4

1 **a** 16 **b** 36 **2** **a** 365 **b** 366 **c** 1461 **3** 8 **4** 32

5 20 **6** **a** 159, 165, 171, 177 **b** 251, 243, 235, 227
c 401, 413, 425, 437

Challenge

Answers will vary, e.g.
There are 9 stars. Each child will get 3 stars.

Activity

12, 42, 18, 48, 60, 36, 24, 54, 30, 48
16, 56, 24, 64, 80, 48, 32, 72, 40, 64

22:1

1 14 **2** 28 **3** 7 **4** 59 **5** 852 **6** 2 **7** 64 **8** 2 **9** 70c

10 1041 **11** 42·36 **12** **a** Friday **b** 4 **c** Monday **d** Tuesday

13 **a** 12, 15, 18, 21, 24 **b** 24, 17, 10, 3 **c** 16, 20, 24, 28, 32

14 **a** **b**

22:2

1 45 **2** 90 **3** 36 **4** 42 **5** 932 **6** 9 m **7** 70 cm **8** 4·0, 4·1
9 8 **10** 1061 **11** **a** Lachlan **b** Sarah **12** **a** 60, 65, 70, 75
b 67, 78, 89, 100 **13** Saturday **14** 57·39

15

	surfaces?	corners?	edges?
cylinder	3	0	0
cube	6	8	12
cone	2	0	0

16 70 cm

17

	3	7	4	5	2	10	6	9
× 9	27	63	36	45	18	90	54	81

Activity

a	6	10	14	18	22	26
b	2	4	8	16	32	64
c	90	81	72	63	54	45
d	11	22	33	44	55	66
e	4	9	14	19	24	29
f	90	82	76	68	60	52

22:3

1 18 **2** 615 **3** **a** B **b** A **4** 73 808
5

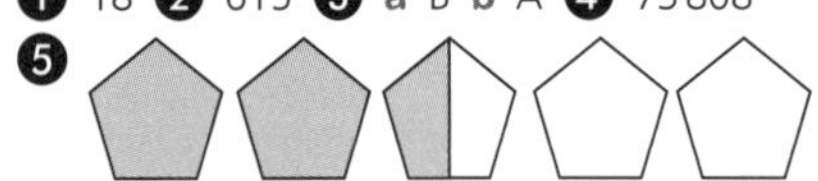

6 55 **7** Answers will vary, e.g. 60, 54, 48, 42, 36, 30.
8 **a** 4 **b** 6 **9** 70 000

22:4

1 **a** Answers will vary, e.g. 2, 4, 8, 16, 32, 64.
b Answers will vary, e.g. 160, 136, 112, 88, 64, 40.
2 **a** 7 **b** 12 **c** 17 **d** 22 **e** 27 **3** 42 **4** **a** $47 **b** $111
5 **a** 48 **b** 96

Challenge

Answers will vary, e.g.
a 25, 50, 75, 100, 125, 150, 175 **b** 90, 60, 30, 0 **c** 9, 18, 27, 36, 45
d 54, 48, 42, 36 **e** 2, 52, 102, 152, 202

Activity

(21) 5 sixths shaded (22) 24 hundredths shaded (23) 3 tenths shaded (24) 0·21 shaded (25) 21 hundredths covered (26) decimal point (27) calendar

23:1

1 14 **2** 28 **3** 28 **4** 49 **5** 813 **6** 0 **7** 21 **8** 7 **9** 16
10 617 **11** Answers will vary, e.g., 7, 10, 13, 16, 19, 22.
12

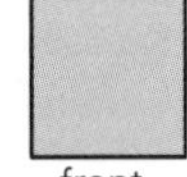

front side

13 6, 12, 8

14 6 **15** yes **16** yes **17** **a** no **b** yes

23:2

1 35 **2** 70 **3** 21 **4** 42 **5** 1021 **6** 7 **7** 49 **8** 56 **9** 28, 35
10 895 **11** Answers will vary, e.g. 100, 200, 300, 400, 500, 600.
12 **a** A **b** B **13** 3, $3\frac{1}{2}$, 4, $4\frac{1}{2}$ **14** **a** 5 **b** 9 **15** yes
16 **a** 30 g **b** 20 kg **17** Tuesday

Activity

18, 0, 36, 24, 6, 30, 12, 42
24, 0, 48, 32, 8, 40, 16, 56

23:3

1 513 **2** 712 **3** **a** 42 **b** 56 **c** 70 **d** 28 **4** A will be circled. **5** C
6 3 × 4, 2 × 6, 14 − 2, 6 × 2, 4 × 3, 6 + 6, and 12 × 1 will be circled.
7 5 **8** **a** 655, 661, 667 **b** 705, 696, 687
9 

23:4

1 26 **2** 60 minutes or 1 hour **3** **a** 64 **b** 30 **c** 24 **4** a cylinder
5 **a** 75 thousands or 75 000 **b** 91 thousands or 91 000
6 **a** $64 **b** $42 **7** 897 mL **8** 647, 538, 429, 320

Challenge

Answers will vary.

Activity

7, 12, 21, 28, 35, 42, 49, 56, 63, 70
4 × 7 = 28
7 × 4 = 28

24:1

1 0 **2** 14 **3** 7 **4** 70 **5** 634 **6** 7 **7** 21 **8** 7 **9** 40c
10 421 **11** cylinder **12** **a** 3 **b** 0 **13** **a** grapes, ham, melon **b** 12 kg
14 yes **15** 52 **16** 14

24:2

1 63 **2** 56 **3** 50 **4** 40 **5** 421 **6** 10 **7** 30 **8** 6·0, 6·1
9 4 **10** 260 **11** C **12** **a** 1400 g **b** 5 kg **c** 500 g **d** 5250 g
13 24 to 5 **14** **a** 4 **b** 6 **15** **a** 104 **b** 156
16 **a** $1\frac{1}{3}$ **b** $2\frac{1}{2}$ **c** $1\frac{2}{10}$ or $1\frac{1}{5}$

Activity

(18) digital watch (19) analog clock (20) 5 × 7 (21) 20 ÷ 5 (22) 18 ÷ 6 (23) counting numbers (24) even numbers (25) odd numbers (26) ordinal numbers (27) digits

24:3

1 624 **2** 274 **3** cylinder, cone, and sphere
4 2 × 12, 20 + 4, 15 + 9, 8 × 3, 12 × 2, 6 × 4, 30 − 6, 29 − 5, 10 + 14, 6 × 4, 3 × 8, 18 + 6, 34 − 10, and 48 ÷ 2 will be circled.
5 true **6** **a** 4:02, 2 past 4 **b** 10:36, 24 to 11
7 834 006 **8** 60 **9** 7·58 **10** **a** 5:06 pm **b** 8:12 am

 • *AUSTRALIAN SIGNPOST MATHS 4 MENTALS* • ISBN 978 0 6557 0884 1

24:4

1 0 or 1 **2**

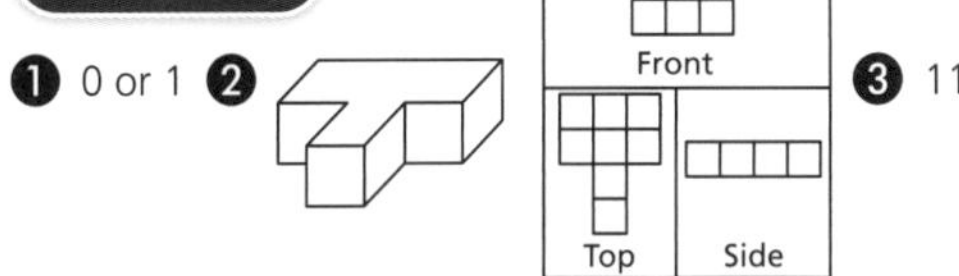

3 11

4 a 2:40 or 20 to 3 b 1:56 or 4 to 2 **5** Sunday

Challenge

Answers may vary, e.g.

This is a 4-digit number. It is made up of 4 tens, 6 units, 8 tenths, and 4 hundredths. It becomes 47 when rounded to the nearest unit.

Activity

35, 63, 42, 21, 56, 14, 49, 70, 28, 0
40, 72, 48, 24, 64, 16, 56, 80, 32, 0

25:1

1 16 **2** 12 **3** 24 **4** 20 **5** 108 **6** 35 **7** 34 **8** 6 **9** 60

10 416 **11** $\frac{9}{4}$, $2\frac{1}{4}$ **12** a 4:39 b 21 minutes to 5 **13** 20 **14** 1 metre

15 a cm b m **16** a m b cm c mm d cm

17 a 36, 40, 44, 48 b 63, 59, 55, 51

25:2

1 21 **2** 42 **3** 32 **4** 64 **5** 446 **6** 8 **7** 4 **8** 18 **9** 19

10 568 **11** yes **12** a 10:37 b 23 to 10 **13** Thursday the 17th

14 537 cm **15** 2 m 67 cm **16** 5 cm **17** 4 m **18** 14 kg

Activity

a 60 b 30 c 15 d 45 e 50 f 10

25:3

1 423 **2** 172 **3** a seven thirteen b <u>13</u> minutes past <u>7</u>

4 a 3:56, 4 to 4 b 6:53, 7 to 7 **5** a 1 cm, 8 mm b 2 cm, 28 mm

6 Estimate: Answers will vary.
Measure: 45 mm

7 a 900 b 1300

8 122

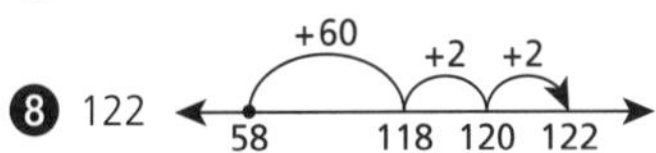

25:4

1 96 **2** Answers will vary, e.g. 8 × 4, 40 − 8, 16 + 16.

3 800 m **4** a 4 cm b 18 cm

Challenge

2 × 7	14
6 × 5	30
5 × 8	40
9 × 7	63
4 × 7	28
6 × 6	36
6 × 8	48
7 × 7	49

8 × 2	16
3 × 7	21
10 × 7	70
6 × 7	42
3 × 8	24
8 × 7	56
7 × 8	56
5 × 7	35

Activity

a 60 b 24 c 60 d 7 e 30 f 2 g 366 h 365

26:1

1 160 **2** 180 **3** 87 **4** 97 **5** 407 **6** 32 **7** 42 **8** 26

9 23 **10** 186 **11** a nine fifty-two b <u>8</u> to <u>10</u> **12** 27 mm

13 51 mm, 39 mm **14** 6 kg **15** 0·79 **16** a 41 b 73 c 165 d 252

17 a 488 b 776

26:2

1 12 **2** 72 **3** 372 **4** 444 **5** 257 **6** 464 **7** 463

8 559 **9** 460 **10** 185 **11** 53 **12** 3 years **13** a 18 cm b 10

14 a Answers will vary. b yes **15** a 13 m^2 b 34 cm^2

16 a 154 b 293 c 401 d 181 **17** a 1300 b 1400

Activity

(1) <u>horizontal</u> line (2) <u>vertical</u> line (3) <u>parallel</u> lines (4) <u>perpendicular</u> lines (5) <u>number</u> line (6) <u>axis</u> of <u>symmetry</u> (19) <u>number</u> expander (20) abacus

26:3

1 368 **2** 348 **3** a 4:02 b 7:51 **4** 28 days

5 Estimate: Answers will vary.
Measure = 5 cm
Measure = 50 mm

6 521 cm **7** a m^2 b cm^2 **8** 145 (+20, +2, +5; 118, 138, 140, 145)

9 a 371 b 676 c 661 d 572

26:4

1 168 **2** a 299 b 560 c 662 d 801 e 592 f 379

3 6 **4** a 37 b 100 **5** 72 **6** a 44 b 25

Challenge

Answers will vary.

Activity

a IV b V c IX d X e XI f XII g II h VIII i 9 j 11 k 6 l 4
m 3 n 8 o 12 p 7

27:1

1 47 **2** 57 **3** 67 **4** 24 **5** 222 **6** 27 **7** 8 **8** 25 **9** 15

10 157 **11** m^2 **12** 122 (+30, +2, +2; 88, 118, 120, 122)

13 a 65 b 43 c 175 d 156 **14** North, East, West, South

15 28 **16** 15, 5, 3 **17** a 70 b 56 c 42

27:2

1 645 **2** 365 **3** 399 **4** 500 **5** 239 **6** 700 **7** 473

8 389 **9** yes **10** 21 **11** a 64 m^2 b 92 cm^2

12 181 (+40, +1, +1; 139, 179, 180, 181)

 • *AUSTRALIAN SIGNPOST MATHS 4 MENTALS* • ISBN 978 0 6557 0884 1

13 a 124 b 116 **14** a 32 b 72 c 152 d 262

15 West **16** 12, 12, 4, 3 **17** 30 **18** 4, 4

Activity

a 52 b 366 c 12 d 10 e 365 f 100 g 24

27:3

1 143 **2** 676 **3** a 899 b 496 **4** a 586 b 525

5 a 799 b 889 **6** a east b west **7** a 6 b Cailin c 30 d 9

8 221 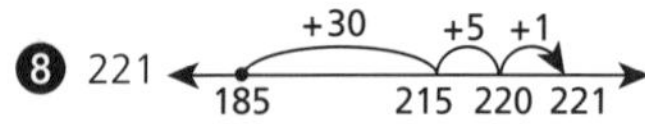

27:4

1 1000, 10 000, 100 000 **2** 232 **3** 8 **4** 32

5 a 1205 b 487 c 621 d 687 **6** 40 **7** a 17 b 60

8 a 24 b 37 **9** $26

Challenge

Answers may vary, e.g.

Deon read the most books. Aimee read the least books. Jean read 2 more books than Lisa.

Activity

a 5 b 15 c 60 d 151 e 165 f 166 g XXV h CX i CCXXI
j CCVII k LXVII l CCCXXXIII

28:1

1 15 **2** 35 **3** 155 **4** 59 **5** 118 **6** 15 **7** 30 **8** 14 **9** 6

10 43 **11** north, west, east, south **12** 23 **13** division sign **14** 6,3

15 2 **16** a 25 b 5 c 15 d 3 **17** 45 640, 45 641, 45 642, 45 643

28:2

1 635 **2** 356 **3** 367 **4** 333 **5** 153 **6** 950 **7** 272 **8** 250

9 125 **10** 163

11 443 (+3, +243; 197, 200, 443)

12 south **13** a 54 b There are 3 more red balloons than yellow. c 27

14 3, 3 **15** Answers will vary.

Activity

a 3 b 5 c 3 d 6 e 7 f 4

28:3

1 853, 854 **2** 430, 431 **3** 10 **4** a 18 b 18 c 24 d 24

5 a 8 b 8 c 3

6 a ÷ b ∟

c 2B d 1B **7** 5

28:4

1 36 **2** 78 **3** 36 **4** 72 **5** 8 km **6** 9 **7** 6 **8** 6

9 11, 11 **10** 588

Challenge

		Total
p	𝍸 𝍸	10
b	𝍸 𝍸 𝍸 𝍸	20

Activity

3, 6, 1, 9, 5, 8, 0, 7, 4, 10
3, 1, 7, 5, 2, 8, 6, 10, 4, 9

29:1

1 83 **2** 62 **3** 65 **4** 19 **5** 177 **6** 182 **7** 658 **8** 783

9 602 **10** 207 **11** 4, 4 **12** 12, 4

13 a grapes b plums c bananas d 2 e 2 out of 6 **14** 456 921

29:2

1 103 **2** 100 **3** 55 **4** 49 **5** 475 **6** 106 **7** 193 **8** 566

9 yes **10** 144 **11** $7\frac{1}{2}$ will be circled. **12** 4, 4 **13** $5 **14** 10

15 a 5A b 3D c 4B d 1C **16** Answers will vary.

Activity

a 3 b 6 c 2 d 4 e 10 f 2 g 5 h 3 i 7 j 7

29:3

1 832, 833 **2** 401, 402 **3** 9 **4** 3 **5** 2 **6** 4, 4

7 dividing (or divide)

8 a The crayons will be shared into 2 groups of 8. b 8 c 8

9 16, 16 **10** 458 297 **11** 18, 6

12 a 8 minutes b 12 minutes c 6 minutes **13** 18 **14** 547 285

29:4

1 14 **2** 19 **3** 9 **4** 324 **5** 23

6 243 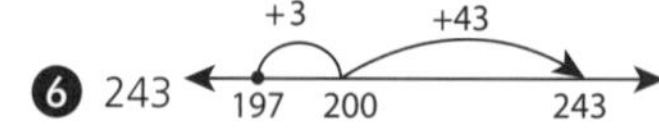

7 56, 168 **8** 423 619 **9** 45 237

Challenge

Answers will vary, e.g.

A group of 5 friends have a bag of 20 chocolates. When the chocolates are shared evenly, each friend will get 4 chocolates. 20 ÷ 5 = 4.

Activity

a WA b WA, NT and Qld c Qld, NSW, Vic and Tas d SA, Vic and Tas.

30:1

1 6 **2** 7 **3** 10 **4** 10 **5** 791 **6** 4 **7** 5 **8** 10 **9** 16

10 208 **11** 8, 4 **12** a 5 rows of 6 = 30 b 30, 30 **13** 6 each, 6

14 a biscuits b second from the left, on the bottom row

15 a 2 b 6

30:2

1 3 **2** 7 **3** 8 **4** 5 **5** 700 **6** 4 **7** 4 **8** yes **9** 3·6, 3·7

10 288 **11** 5, 5 **12** 9 **13** $\frac{7}{4}$ will be circled. **14** Even will be circled.

15 **a** Saturday **b** 5 **c** 8/12/2027 **d** 24 **16** **a** 4:55 pm **b** 5:59 am

Activity

1, 4, 7, 2, 9, 6, 3, 8, 5, 10
7, 4, 9, 0, 8, 2, 5, 10, 3, 6

30:3

1 195 **2** $139 **3** 2 **4** 4 **5** 6 **6** 15, 5

7 **a** 20 minutes **b** 40 minutes **8** **a** 5 **b** 10 **9** $\frac{5}{2}$ will be circled.

10 **a** 4:05 **b** 5 past 4 **11** 4:24 pm **12** **a** $1.65 **b** $1.30

30:4

1 **a** 2, 6, 8 **b** 5 (The totals that occur twice are 17, 15, 19, 18 and 20.)

2 7 **3** 24 **4** **a** 32 **b** 40

Challenge

a

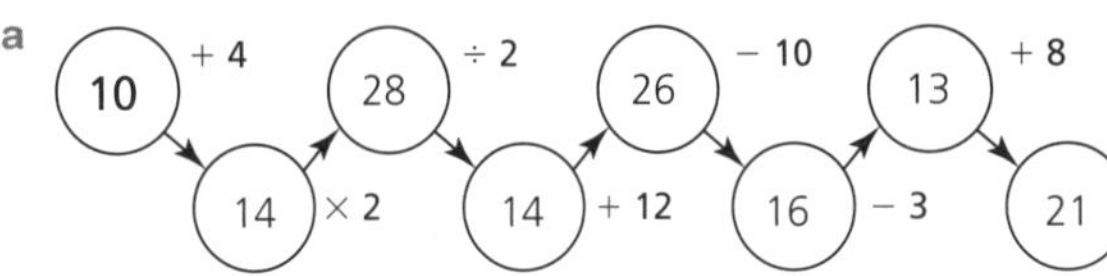

b Answers will vary.

Activity

6, 42, 30, 12, 48, 0, 18, 36, 54, 24
7, 49, 35, 14, 56, 0, 21, 42, 63, 28

31:1

1 5 **2** 5 **3** 3 **4** 3 **5** 229 **6** 4 **7** 4 **8** 5 **9** 6

10 775 **11** **a** $\frac{2}{4}$ Answers may vary. **b** $\frac{2}{6}$ Answers may vary.

12 Odd will be circled. **13** **a** 9 **b** 9 **14** **a** 6:25 **b** 25 past 6

15 47 minutes **16** 4, 4 **17** 25 **18** 7

31:2

1 6 **2** 6 **3** 9 **4** 9 **5** 35 **6** 8 **7** 8 **8** 9 **9** 5

10 402 **11** 22 **12** Even will be circled. **13** **a** 4 **b** 2nd of September

c Tuesday **d** 48 **14** **a** **b**

Activity

×	3	7	6	9	8	4
4	12	28	24	36	32	16
6	18	42	36	54	48	24
7	21	49	42	63	56	28
8	24	56	48	72	64	32
9	27	63	54	81	72	36

31:3

1 8 **2** 5 **3** 4 **4** 4 **5** 8 **6** 8 **7** Even will be circled.

8 5:05 **9** 34 minutes **10** **a** 9:25 **b** 25 past 9

11 6 rows of 7 = 42, 7, 7 **12** 59 249

13 **a** $\frac{2}{2}$ Answers may vary. **b** $\frac{6}{10}$ Answers may vary. **14** **a** 6 **b** 9

15 15 (number line: 61 −40 → 21 −1 → 20 −5 → 15)

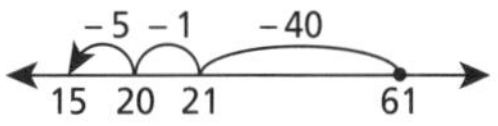

31:4

1 1 h 54 min **2** $16 **3** 11 **4** **a** VIII **b** IX **5** $26 **6** **a** 74 **b** 92

Challenge

Answers will vary.

Activity

a 8 **b** 6 **c** 4 **d** 12 **e** 2 **f** 3 **g** 4 with 4 left over

32:1

1 10 **2** 20 **3** 70 **4** 730 **5** 32 **6** 24 **7** 40 **8** 24 **9** 24

10 691 **11** **a** 30 minutes **b** 45 minutes **12** 3 rows of 5 = 15, 5, 5

13 yes **14** 40 **15** true **16** **a** 2 **b** 3 **17** true **18** yes

32:2

1 6 **2** 11 **3** 7 **4** 5 **5** 588 **6** 1200 **7** 180 **8** 280

9 3500 **10** 557 **11** 50c, 10c, 10c, 5c **12** 5 **13** 60

14 72 **15** **a** 6, 9 **b** yes **16** true **17** **a** 20 **b** 6 **18** yes

19 **a** 6000 grams **b** 7 kg **c** 0·4 kg

Activity

a 42, 6, 7 **b** 28, 7, 4 **c** 48, 8, 6 **d** 54, 6, 9 **e** 72, 9, 8 **f** 56, 7, 8

32:3

1 155 **2** 1020 **3** 912 **4** 6 **5** 7 **6** 5

7 **a** 3 rows of 6 = 18 **b** 6, 6 **8** 35 **9** **a** 15, 20 **b** yes **10** 11

11 **a** 50 **b** 15 **c** 65 **12** **a** Estimates will vary. (8 L)

b Estimates will vary. (4 L, 5 mL) **13** **a** 9 kg **b** 4 kg

32:4

1 92 **2** 138 **3** **a** 16 cm^2 **b** 8 cm^2 **c** 4 cm^2 **4** yes

5 10 cm **6** 934 823 **7** 103

Challenge

Answers will vary, e.g.
2 × 6 = 12, so 2 × 12 = 24

a 3 m **b** 12 m **c** 30 cm **d** 3 cm **e** 10 cm **f** 12 m **g** 12 m **h** 2 m

33:1

1 12 **2** 24 **3** 20 **4** 40 **5** 145 **6** 20 **7** 21 **8** 30 **9** 42

10 578 **11** 27 **12** **a** 12, 18 **b** yes

⑬ 0·5 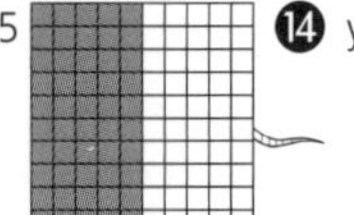 ⑭ yes ⑮ Estimates will vary. (17)

⑯ **a** 6 kg **b** 46 g ⑰ 24 ⑱ **a** 1 **b** 4 **c** 1 **d** 4

33:2

❶ 15 ❷ 8 ❸ 20 ❹ 10 ❺ 842 ❻ 112 ❼ 112

❽ 126 ❾ 126 ❿ 306 ⓫ 9, 9 ⓬ 5 ⓭ 32

⓮ Explanations will vary. $(3 \times 10) + (3 \times 10) + (3 \times 7) = 81$

⓯ Rectangle: Estimates will vary, $A = 12$ cm^2 Triangle: $A = 6$ cm^2

⓰ cold

Activity

30, 54, 36, 18, 48, 12, 42, 60, 24, 0
45, 81, 54, 27, 72, 18, 63, 90, 36, 0

33:3

❶ 977 ❷ 502 ❸ 405 ❹ 8 ❺ 2 ❻ 4 ❼ 42

❽ **a** 6 cm^2 **b** 3 cm^2 ❾ **a** 2 **b** 6 ❿ D, C, A, B ⓫ **a** 0·7 kg **b** 0·4 kg

⓬ Answers will vary. You could put people into groups of 2, 3, 4, 6, 8, or 12. ⓭ 4

33:4

❶ 6570 ❷ 567 ❸ 1 kg ❹ Answers may vary. (true) ❺ 3 ❻ 385

❼ **a** 450 **b** 1200 ❽ **a** 27 **b** 81 ❾ 40 ❿ 72 ⓫ 210

Challenge

Answers will vary.

Activity

(11) <u>acute</u> angle (12) <u>right</u> angle (13) <u>straight</u> angle (14) <u>obtuse</u> angle
(15) <u>reflex</u> angle (16) revolution (17) <u>vertex</u> of an angle
(18) <u>arm</u> of an angle

34:1

❶ 152 ❷ 91 ❸ 40 ❹ 5 ❺ 36 ❻ 3·5, 3·6 ❼ \$7.50

❽ \$10.20 ❾ \$20.70 ❿ 107 ⓫ **a** 28 g **b** 13 kg

⓬ **a** 4 **b** 3 ⓭ **a** 7, 7 **b** 8, 8 ⓮ 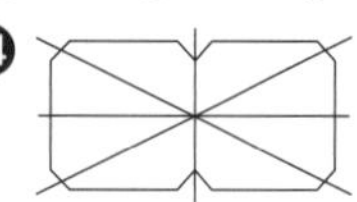

⓯ A ⓰ Answers may vary. \$5 + \$2 + 20c

34:2

❶ 402 ❷ 510 ❸ 60 ❹ 50 ❺ 221 ❻ 9·3, 9·4

❼ \$70.50 ❽ \$21 ❾ \$3.55 ❿ 941 ⓫ 8, 8

⓬ Explanations will vary, e.g. $(7 \times 10) + (7 \times 10) + (7 \times 2) = 154$

⓭ **a** C **b** A ⓮ \$6.45

⓯ Answers will vary. You could put people into groups of 2, 4, 8, or 16.

Activity

0, 6, 3, 9, 1, 5, 7, 2, 8, 4
2, 5, 1, 9, 7, 4, 10, 8, 3, 6

34:3

❶ 977 ❷ 502 ❸ 405 ❹ 8, 8

❺ Explanations may vary, e.g. $(8 \times 10) + (8 \times 10) + (8 \times 10) + (8 \times 10) + (8 \times 7) = 376$

❻ B ❼ **a** 378 g **b** 92 kg ❽ **a** 7, 7 **b** 7, 7

❾ \$10 + \$5 + 50c + 20c + 10c + 5c ❿ \$22.50

34:4

❶ 600 ❷ 4 will be circled. ❸ 6 ❹ 14 ❺ \$16.55

❻ **a** 197 **b** 803

Challenge

Answers will vary, e.g.

49 is an odd number. It becomes 50 when rounded to the nearest ten.
98 is double this number $7 \times 7 = 49$, $60 - 11 = 49$.

Activity

a 50 cents **b** 95 cents **c** 65c **d** 20c **e** 90c **f** \$1 **g** 70c **h** 95c
i 25c **j** 55c

35:1

❶ 50 ❷ 90 ❸ 8 ❹ 17 ❺ 17 ❻ 500 ❼ 3 ❽ 9 ❾ 6000

❿ 752 ⓫ Answers may vary. \$1 + \$1 + 20c + 5c

⓬ 20c + 10c + \$2 + \$1 = \$3.30 ⓭ Jon ⓮ **a** yes **b** yes

⓯ **a** yes **b** no ⓰ **a** 0·5 **b** 0·6

35:2

❶ 300 ❷ 800 ❸ 4 ❹ 60 ❺ 125 ❻ 9400 ❼ 3500

❽ 5300 ❾ 39 000 ❿ 274

⓫ 

⓬ **a** \$15.22 **b** \$15.20 ⓭ 10c + \$2 + \$2 + \$10 = \$14.10

⓮ The rectangle will be circled.

⓯ **a** 60 000 **b** 35 **c** 3000 **d** 2600

Activity

^{1}3	0	^{2}5	■	^{3}1	■	^{4}7
4	■	2	■	^{5}5	0	0
^{6}8	^{7}1	4	9	■	■	4
■	^{8}2	4	■	^{9}7	1	1
■	9	■	■	0	■	■
■	^{10}2	0	1	8	■	■

35:3

1 8 **2** 5 **3** 4 **4** 4 **5** 8 **6** 8 **7** 50c + $5 + $20 = $25.50

8 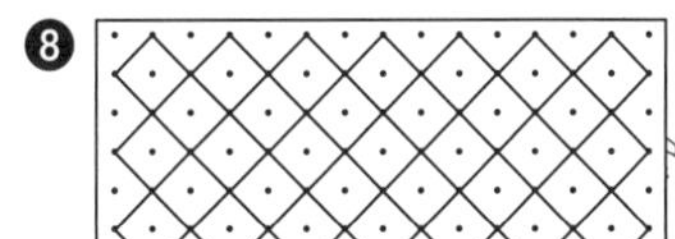 **9** **a** yes **b** yes

10 **a** 2000 **b** 49 **c** 9000 **d** 1910 **11** 6300 **12** **a** $17.25 **b** $12.80

35:4

1 **a** 14 **b** 104 **2** triangles, squares, and hexagons **3** 7

4 **a** 85 763 **b** 92 **c** 7453 **d** 5760 **5** $240 **6** 210 **7** $76.40

Challenge

Answers will vary.

Activity

1, 4, 7, 2, 9, 6, 3, 8, 5, 10
7, 4, 9, 0, 8, 2, 5, 10, 3, 6

36:1

1 60 **2** 900 **3** 300 **4** 80 **5** 19 **6** 9 **7** 4 **8** 30 **9** 7

10 8 **11** **a** A, C, E, F, G **b** D and B **c** H and I **d** D, B, H, I

12 **a** 5, 5 **b** 4, 40 **13** 40 **14** 10 **15** **a** 800 **b** 500 **c** 200

16 **a** 40 **b** 20 **c** 32

17 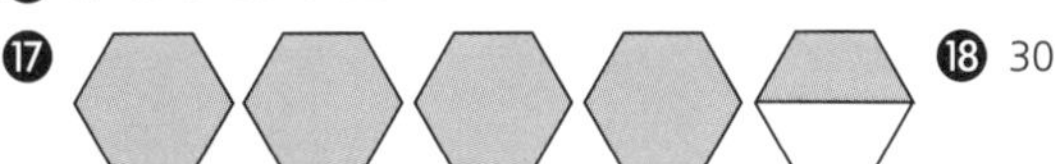 **18** 30

36:2

1 36 470 **2** 84 020 **3** 43 500 **4** 89 400 **5** 1155

6 8463 **7** 7830 **8** 638 **9** 200 000 **10** 335 **11** **a** C **b** B

12 ◇ **13** **a** 6700 **b** 88 **c** 5000 **d** 2500

14 yes **15** $12 600 **16** **a** 26 **b** 15 **17** **a** 13 **b** 5 **c** 9 **d** 9 **e** 8 **f** 5

Activity

a 2 **b** 4 **c** 6 **d** 12 **e** 18 **f** 38

36:3

1 324 **2** 388 **3** $350 **4** **a** 27 **b** 13

5 **a** 453 + 97 = 453 + 100 − <u>3</u> = <u>550</u>

b 637 + 95 = 637 + 100 − <u>5</u> = <u>732</u>

c 254 + 92 = 254 + 100 − <u>8</u> = <u>346</u> **6** 5·4 m

7 **a** <u>3675</u> mm or <u>3</u> m <u>675</u> mm **b** <u>1650</u> mm or <u>1</u> m <u>650</u> mm

36:4

1 1924 **2** **a** true **b** true **3** X **4** yes **5** 6

6 **a** 56 **b** 49 **7** 8

Challenge

Answers will vary.

Activity

a $1.35 **b** $1.30 **c** $2.50 **d** $8.15 **e** $6.00 **f** $2.00

37:1

1 4 **2** 11 **3** 6 **4** 8 **5** 19 **6** 19 **7** 20 **8** 8 **9** 5

10 928 **11** **a** 8 **b** 6 **12** **a** 25 **b** 35 **13** **a** 38 **b** 37

14 7 km **15** data **16** **a** 42 **b** 51 **17** Answers will vary. **18** 56

37:2

1 7 **2** 30 **3** 11 **4** 7 **5** 18 **6** 7 **7** 9 **8** 8 **9** 12

10 942 **11** **a** 32 **b** 14

12 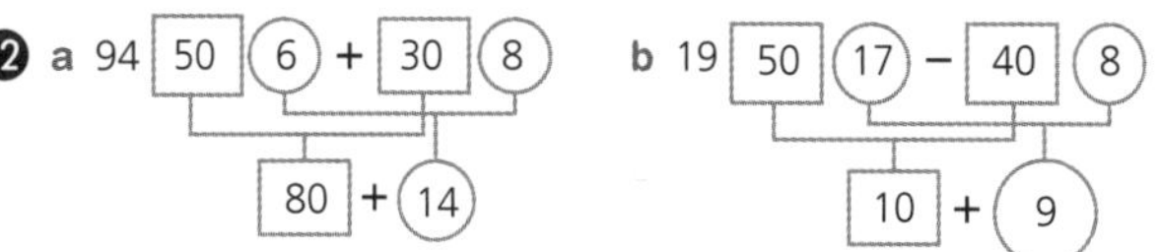

13 **a** no **b** yes **c** A, D, and H **d** B, E, G and I **14** $44.70

Activity

(1) metres (2) centimetres (3) millimetres (4) square metres
(5) square centimetres (6) litres (7) millilitres (8) kilograms

37:3

1 324 **2** 388 **3** **a** 67 **b** 47 **4** **a** 7 **b** 7 **c** 8 **d** 6

5 **a** cube **b** square **c** 4A **d** 1B **6** **a** Monday **b** 25 **c** 100

37:4

1 **a** 486, 456, 447 **b** 523, 513, 505 **2** **a** 16 **b** 15

3 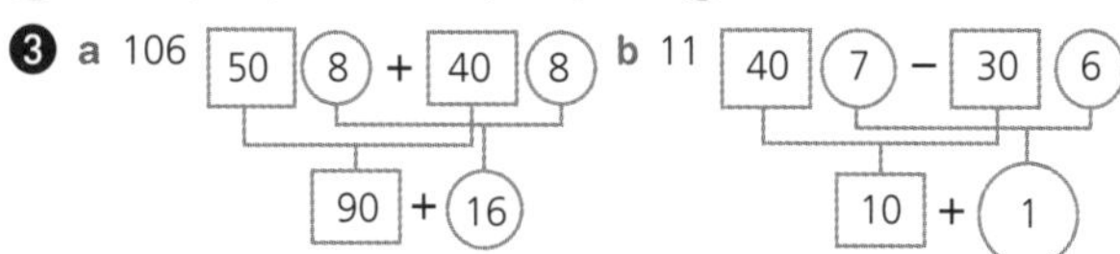

4 **a** 586 + 99 = 586 + 100 − <u>1</u> = <u>685</u>

b 293 + 94 = 293 + 100 − <u>6</u> = <u>387</u> **5** 14 **6** 8

Challenge

Answers will vary, e.g.

The decimal is made up of 2 hundreds, 6 tens, 1 unit, 7 tenths, and 3 hundredths. It becomes 262 when rounded to the nearest unit.

Activity

Answers will vary.

How would you name the lizards?

Find two lizards hidden on each page of this book.

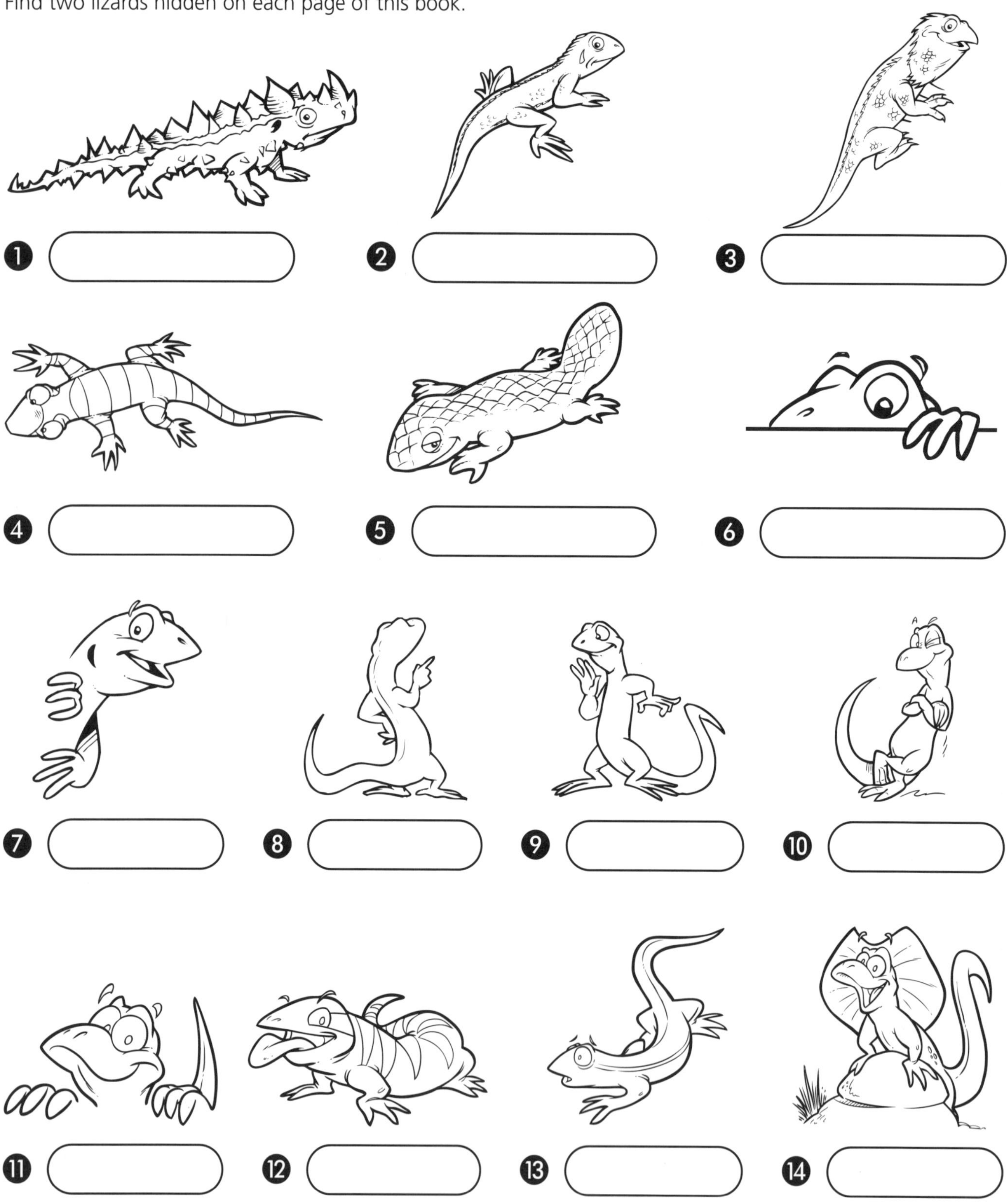

Real names:

1 Horned lizard **2–3** Bearded dragon **4** Gecko **5** Shingle-back lizard

6–11 Common garden lizard **12** Blue-tongue lizard **13** Skink **14** Frilled-neck lizard

17:3 ☐ out of 7

1. True (T) or false (F)? To multiply by 6 you can multiply by 3, then double. ______

2. Write the decimal shown by this model.

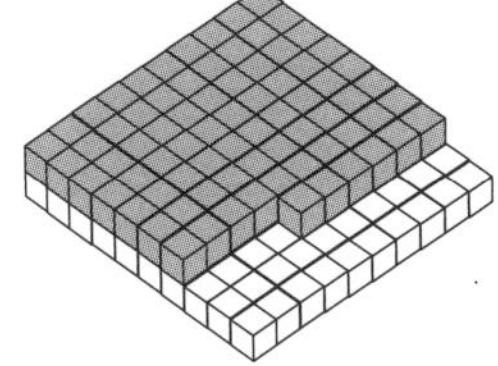

______

3. How many litres in:
   a 4000 mL? ______ **b** 9000 mL? ______

4. Would you use litres or millilitres to measure a:
   a mug? ______ **b** bucket? ______

5.

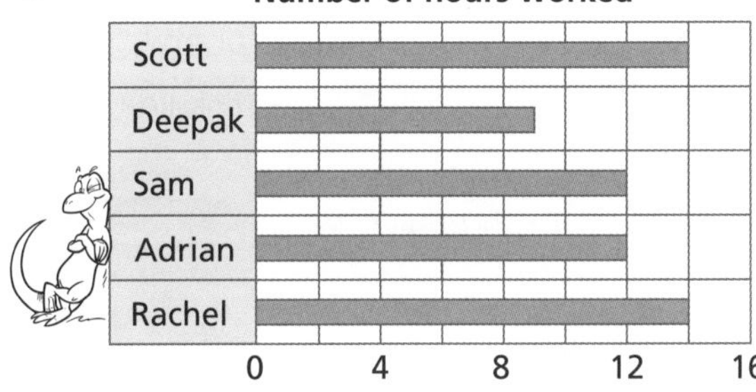

   a For how long did Adrian work? ______
   b How many hours were worked altogether? ______

6. **a** Colour the water level, if 700 mL of water is put in this container.
   b How much water would be left if 100 mL was poured out? ______ mL

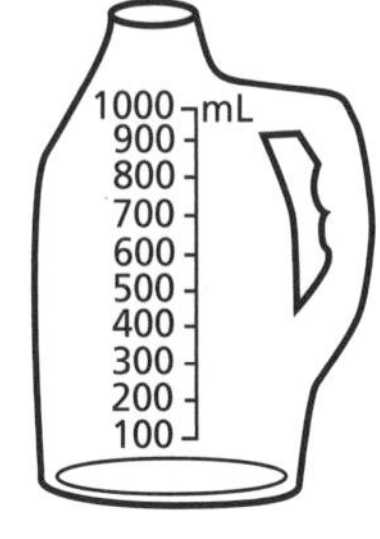

7. Write the short form for:
   a 8 metres ______ **b** 74 millilitres ______
   c 92 litres ______ **d** 35 millimetres ______

17:4 Extension ☐ out of 9

1. I opened a book. The sum of the two page numbers I saw was 73. What were the numbers?
   ______ and ______

2. I have four boxes of matches. 2 boxes have 9 matches left in each. The other boxes have 12 matches in each.

   How many matches do I have? ______

3. Does $64 \times 3 = 3 \times 64$? ______

4. $4 \times 3 \times 20$ ______

5. If $23 \times 11 = 253$, what does 22×11 equal? ______

6. How many 5c cent coins have the same value as $2? ______

7. What is the total value of sixteen 5c coins and ten 20c coins? ______

8. How many 10c coins have the same value as two hundred 5c coins? ______

9. How many days in autumn? ______

List the name and capacity of containers in your kitchen or bathroom that are measured in L or mL.

______
______
______
______

What could this graph about living things represent?

______

Explain what you think each column could stand for.

______
______
______

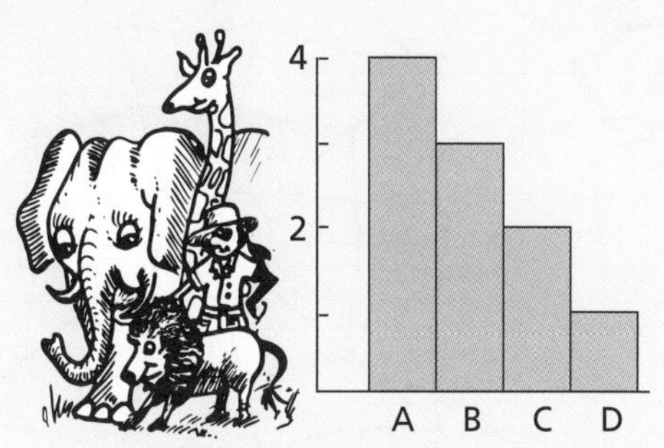

18:1 ☐ out of 17

1. 1×3 ______
2. 1×6 ______
3. 2×3 ______
4. 2×6 ______
5. $\begin{array}{r} 45 \\ -27 \\ \hline \end{array}$
6. $54 + 20$ ______
7. $23 + 20$ ______
8. $35 +$ ______ $= 40$
9. $42 +$ ______ $= 50$
10. $\begin{array}{r} 53 \\ -27 \\ \hline \end{array}$
11. Colour 3 tenths of this hundreds block and write the decimal.

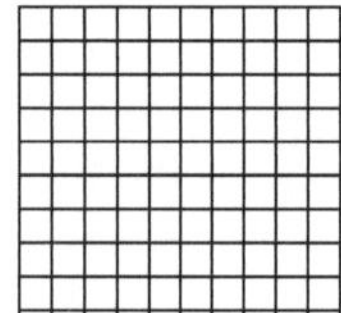

___ . ___

12. Would it take about 1 L, 1 800 mL, 600 mL or 20 L of water to fill 2 mugs?

______

13. 1000 mL = ______ L
14. Where could you find a right angle in your house? ______
15. 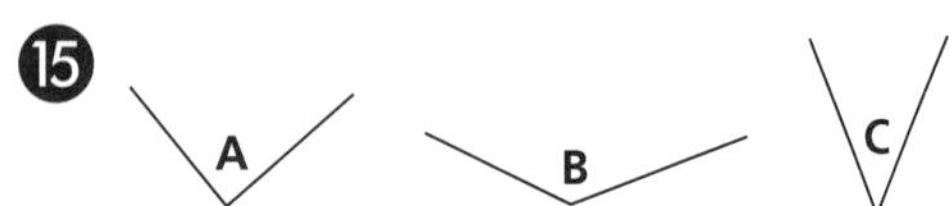

Which angle is the right angle? ______

16. $200\,000 + 30\,000 + 1000 + 600 + 20 + 9$

______

17. **a** Minutes in half an hour. ______

b Minutes in one hour. ______

18:2 ☐ out of 16

1. 3×3 ______
2. 6×3 ______
3. 10×3 ______
4. 10×6 ______
5. $\begin{array}{r} 46 \\ -18 \\ \hline \end{array}$
6. $8 \times 3 + 3$ ______
7. 9×3 ______
8. Halve 44. ______
9. Double 43. ______
10. $\begin{array}{r} 73 \\ -58 \\ \hline \end{array}$

11. 42 birds. 26 flew away. How many are left? ______
12. Would you use L or mL to measure the capacity of a:

a bucket? ______ **b** tablespoon? ______

c bath tub? ______ **d** glue stick lid? ______

13. What unit of measurement has been left off this label?

______

14. What is the total quantity held?

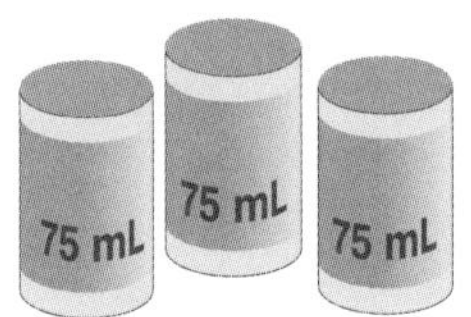

______

15. How many:

a quarter turns make a full turn? ______

b right angles make a full turn? ______

c right angles make a straight angle? ______

16. Minutes in half an hour? ______ min

Minutes in a quarter of an hour? ______ min

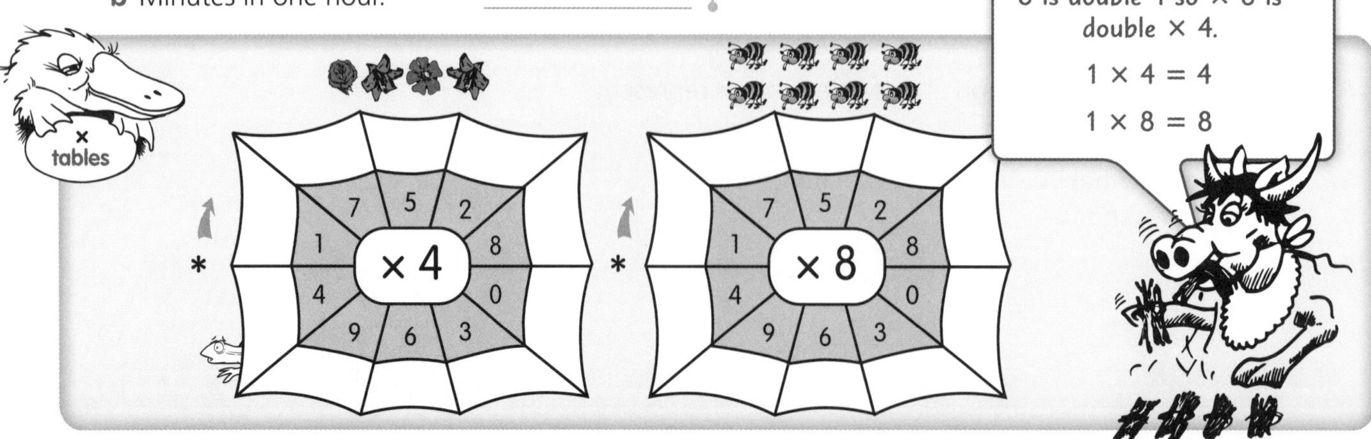

18:3 ☐ out of 7

1

tens	ones
4	3
− 1	6

2

tens	ones
6	5
− 4	8

3   

60 mL 50 mL 75 mL

How many mL altogether? ______

4 Use the short form to write:

a 500 millilitres ______

b 80 litres ______

c 1·5 metres ______

5 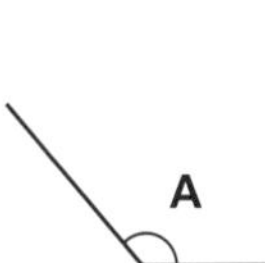  

a Write the angles in order of size with the smallest first. ______

b Which angle is a right angle? ______

6 **a** The time on clock **A** is ______.

b Show 10 past 12 on clock **B**.

c Is the smaller angle between the hands on clock **A** larger than the angle on clock **B**?

______

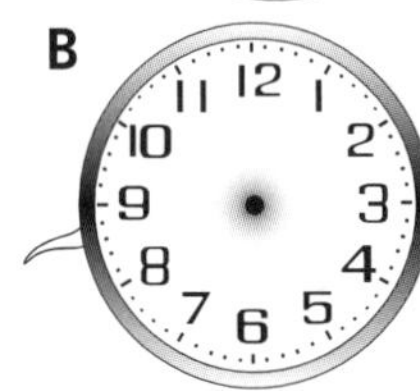

7 37 is ______ more than 21.

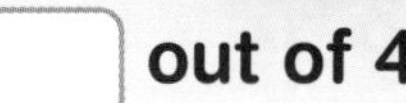

18:4 ☐ out of 4

1 Shelby bought four 2000 mL containers of milk. How many litres did she buy altogether? ______

2 **a** The hour hand is pointing at 9. The minute hand and hour hand make a right angle. What time is it? ______

b Is it possible for the hour hand to point at 6 and be at right angles to the minute hand? ______

3 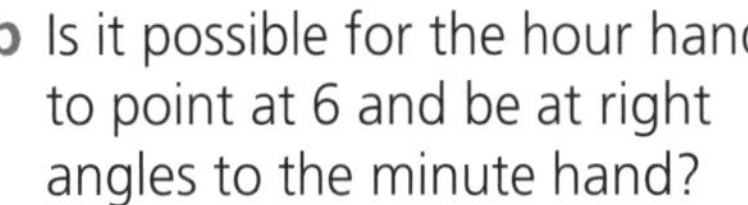

How many students could be given:

a 2 pens? ______ **b** 3 pens? ______

4 If '<' means 'is less than', write true or false for:

a $2 \times 8 < 5 \times 4$ ______

b $35 < 3 \times 10$ ______

c $3 \times 9 < 7 \times 4$ ______

Challenge

Complete each column as quickly as you can. Once finished, check your answers and record your time.

2 × 6	______	8 × 2	______
2 × 3	______	4 × 2	______
5 × 3	______	3 × 3	______
0 × 6	______	3 × 6	______
5 × 6	______	10 × 6	______
1 × 6	______	4 × 3	______

Time = ______ seconds

 • *AUSTRALIAN SIGNPOST MATHS 4 MENTALS* • ISBN 978 0 6557 0884 1

19:1 ☐ out of 18

1. 1 × 9 ______
2. 15 + 9 ______
3. 2 × 9 ______
4. 18 − 13 ______
5. $\begin{array}{r} 43 \\ -16 \\ \hline \end{array}$
6. Double 24. ______
7. Half 42 ______
8. Add $13 and $8. ______
9. 20 divided by 10 ______
10. $\begin{array}{r} 54 \\ -37 \\ \hline \end{array}$

11. A  B  C 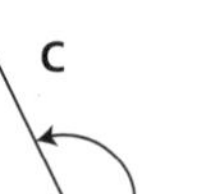 D 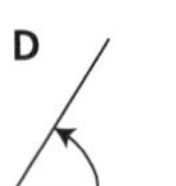

    Put these angles in order from smallest to largest. ____________

12. 

    a 3 rows of 9. ______ **b** 5 rows of 9. ______

13. Draw 4 groups of 6.

14. If there are 10 dots in each row, how many dots are in 6 rows? ______
15. Is this a right angle? ______
16. 5 × 3 = 15 so 5 × 6 = ______
17. True or false? When multiplying by 9 you could multiply by 10 and subtract the original number. ______
18. 9, 18, 27, ______, ______, ______, ______, ______

19:2 ☐ out of 18

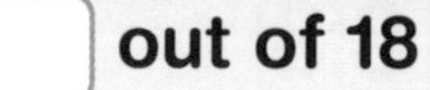

1. 4 × 9 ______
2. 6 × 9 ______
3. 3 × 9 ______
4. 8 × 9 ______
5. $\begin{array}{r} 85 \\ -66 \\ \hline \end{array}$
6. Multiply 5 by 9. ______
7. 7 times 9 ______
8. 18 divided by 3 ______
9. Sum of 19 and 17 ______
10. $\begin{array}{r} 92 \\ -57 \\ \hline \end{array}$

11. The decimal covered is:

    a ______ tenths

    b ______ hundredths

    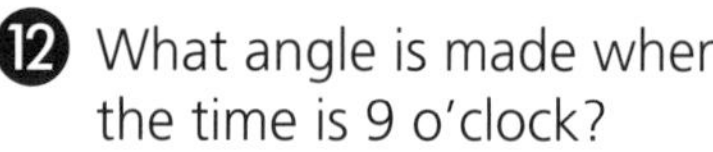

12. What angle is made when the time is 9 o'clock? ______
13. How many legs on 8 horses? ______
14. 4 × 3 = 12 so 4 × 6 = ______
15. 3 children have 4 birds each. How many birds altogether? ______
16. × 5 answers end in ______ or ______.
17. 6 × 11 = ______

    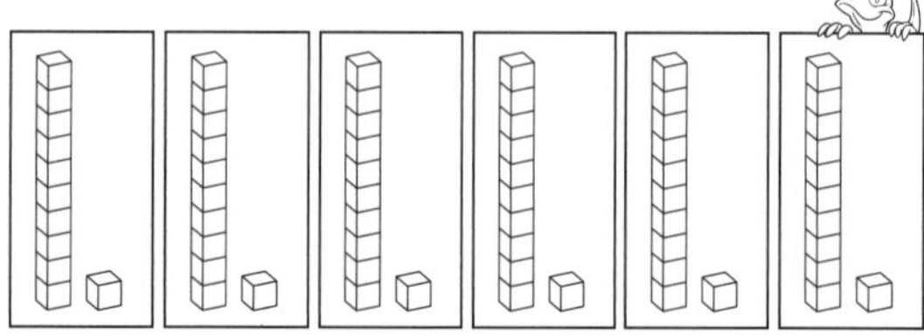

18. **a** Draw each diagonal that can be drawn from the dot on this hexagon.

    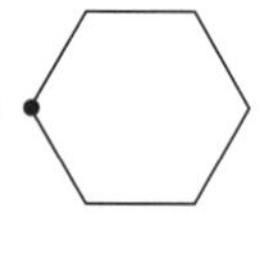

    b How many triangles have you made in the hexagon? ______

Finish these to find the pattern.

1 × 9 = [0] [9], [0] + [9] = ______	6 × 9 = [ ] [ ], [ ] + [ ] = ______
2 × 9 = [1] [8], [1] + [8] = ______	7 × 9 = [ ] [ ], [ ] + [ ] = ______
3 × 9 = [2] [7], [2] + [7] = ______	8 × 9 = [ ] [ ], [ ] + [ ] = ______
4 × 9 = [ ] [ ], [ ] + [ ] = ______	9 × 9 = [ ] [ ], [ ] + [ ] = ______
5 × 9 = [ ] [ ], [ ] + [ ] = ______	10 × 9 = [ ] [ ], [ ] + [ ] = ______

Nine times tables:
The digits in each answer always add up to ______.

19:3 ☐ out of 9

1.

tens	ones
6	2
− 2	8

2.

tens	ones
7	4
− 5	7

3. What angle is made when the time is 3 o'clock? ______

4. How many sides on:
   a 7 hexagons? ______ b 8 octagons? ______

5. a Draw each diagonal that can be drawn from the dot on this pentagon.
   b How many triangles have you made in the pentagon? ______

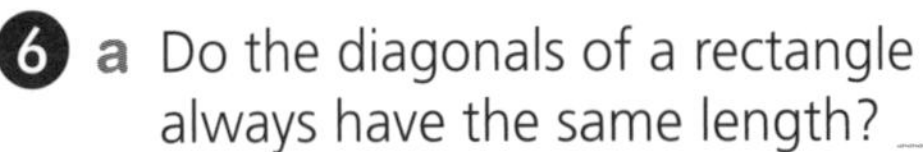

6. a Do the diagonals of a rectangle always have the same length? ______
   b Do they cut each other in half? ______

7. Circle the shape below that can be made from these three triangles.

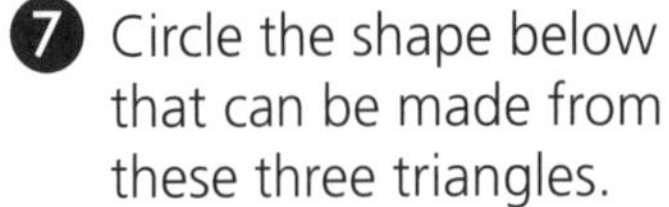

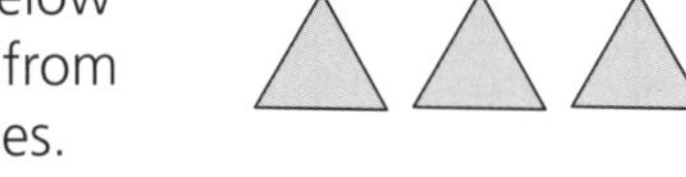

A 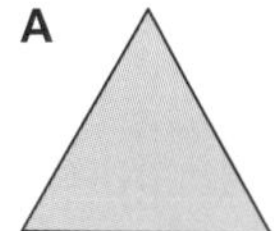 B C 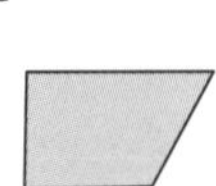

8. a 5 students read 3 books each. How many books were read? ______

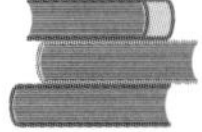

   b 20 more books were then read. How many books were read altogether? ______

9. a Draw 3 axes of symmetry for this triangle.
   b Do all triangles have 3 axes of symmetry? ______

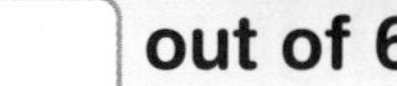

19:4 ☐ out of 6

1. Which is larger, an angle of a square or of a regular hexagon? ______

2. Complete inside the brackets first to find:
   a $(3 \times 12) + (3 \times 12)$ ______
   b $(4 \times 6) + (8 \times 8)$ ______

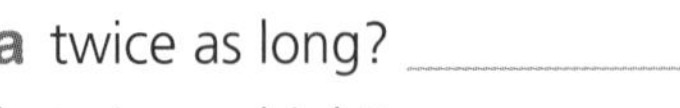

3. 6 stamps fit into this rectangle. How many stamps would fit into a rectangle that is:
   a twice as long? ______
   b twice as high? ______

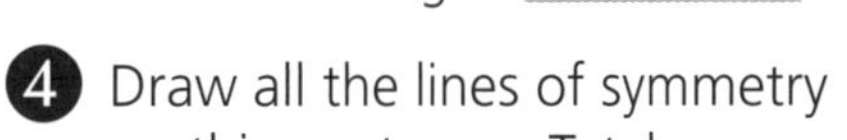

4. Draw all the lines of symmetry on this pentagon. Total = ______

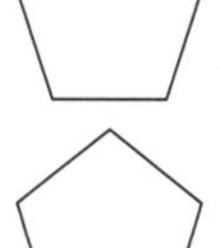

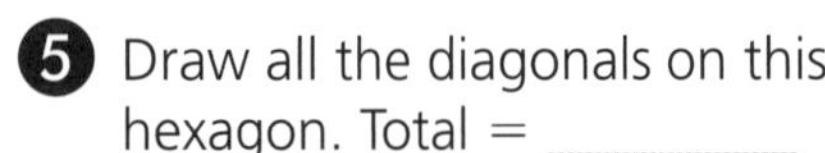

5. Draw all the diagonals on this hexagon. Total = ______

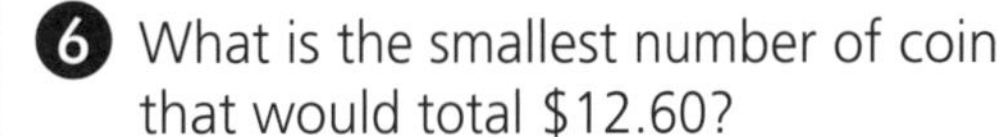

6. What is the smallest number of coins that would total $12.60? ______

Draw as many different composite shapes as you can using two squares and two triangles.

Example:

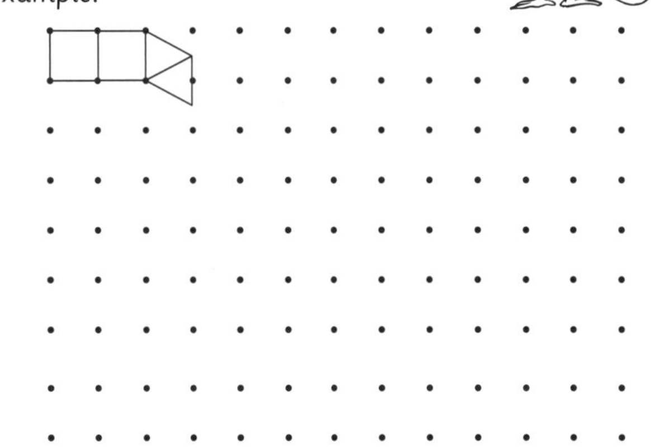

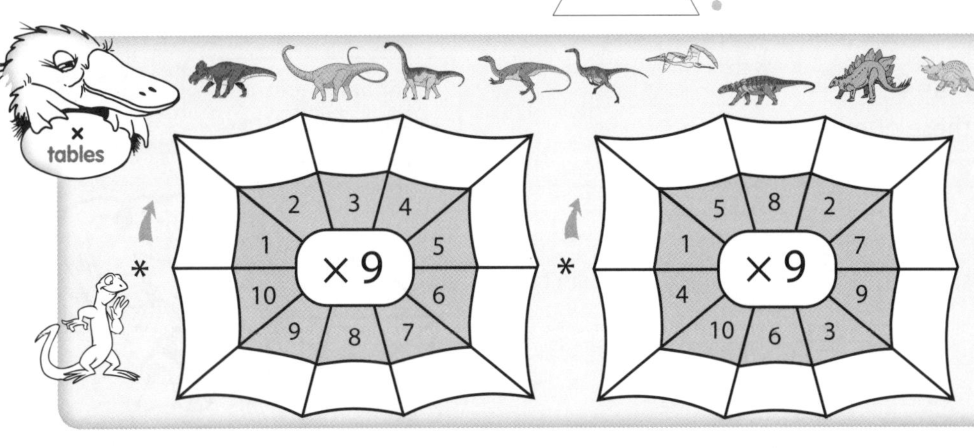

Make sure you learn these tables.

The digits of your answers will add to make 9.

© PEARSON AUSTRALIA 2024 • *AUSTRALIAN SIGNPOST MATHS 4 MENTALS* • ISBN 978 0 6557 0884 1

20:1 out of 18

1. 0 × 9 ______
2. 23 − 6 ______
3. 10 × 9 ______
4. 19 + 8 ______
5. $\begin{array}{r} 294 \\ +157 \\ \hline \end{array}$
6. 8 shared by 2 ______
7. 10 ÷ 2 ______
8. 10 times 7 ______
9. 12 divided by 3 ______
10. $\begin{array}{r} 537 \\ +396 \\ \hline \end{array}$

11. a 4 rows of 5 ______
    b 4 rows of 7 ______
12. Use the picture above to find how many groups of 7 are in 28. ______
13. × 10 answers always end in ______.
14. I can read 8 books a month. How many books can I read in 6 months? ______

15. What date is the:
    a third Friday? ______
    b second Sunday? ______

DECEMBER

Sun	Mon	Tue	Wed	Thu	Fri	Sat
		1	2	3	4	5
6	7	8	9	10	11	12
13	14	15	16	17	18	19
20	21	22	23	24	25	26
27	28	29	30	31		

16. Write six thousand, three hundred. ______
17. 56 000 + 400 + 50 + 0 ______

18.    

The total value of these coins in decimal form. $______

20:2 out of 15

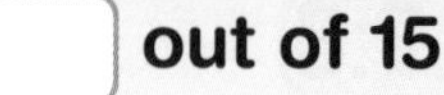

1. 3 × 8 ______
2. 6 × 8 ______
3. 16 ÷ 4 ______
4. 20 ÷ 5 ______
5. $\begin{array}{r} 683 \\ +274 \\ \hline \end{array}$
6. 0·3, 0·4, ______, ______
7. 8 multiplied by 6 ______
8. 3 times 9 ______
9. 18 divided by 6 ______
10. $\begin{array}{r} 643 \\ +179 \\ \hline \end{array}$
11. How many sides on:
    a 3 hexagons? ______ b 6 octagons? ______
12. Jake gave $663 to charity this year and $345 last year. How much did he give altogether? ______
13. 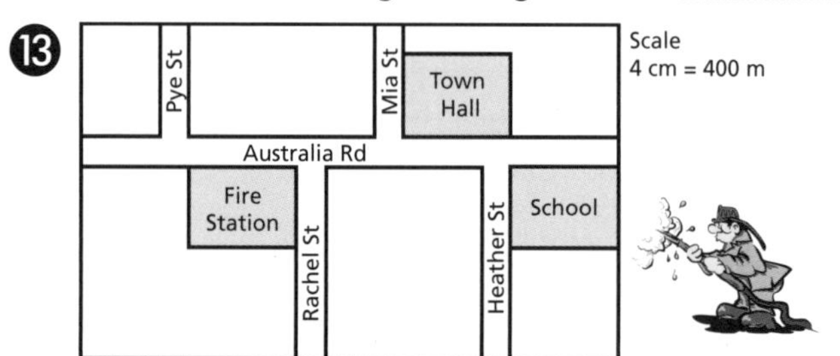

    a What is on Mia St? ______
    b What is on Rachel St? ______
    c The scale is: 1 cm = ______
    d How far is it from the fire station to the school? ______
14. Colour $3\frac{1}{2}$ pentagons

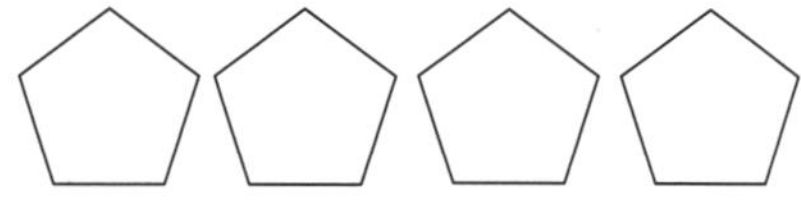

15. 

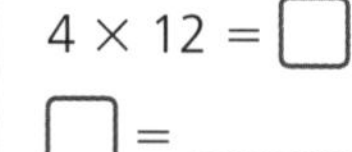

4 × 12 = ☐
☐ = ______

Turn to ID card C on page 8.
Give the answers for these numbers.

(16) ______ (17) ______
(18) ______ (19) ______
(20) ______ surface (21) ______ surface
(22) ______ (23) ______

20:3 ☐ out of 13

1

hund.	tens	ones
2	0	8
+ 1	9	3

2

tens	ones
$5	0
+$1	8

3 4 photo frames can fit along the length of my shelf. How many could fit if the shelf was:

a two times as long? ________

b six times as long? ________

4 The value of these coins.  ________

5 How many 20c coins could I take from:

a $1? ________ **b** $2? ________

c $3? ________ **d** $10? ________

6 20 balls shared between 2 classes. ________each

7 57 000 + 400 + 90 + 2 ________

8 How many tens could I get from 6982?________

9 Write the second smallest number possible using the digits 5, 3, 7, 2, 9 and 4. ________

10 On this rectangular prism, how many:

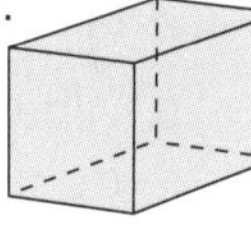

a faces? ________

b edges? ________

11 Write 349 in words.________

____

12 This year Josie sold 528 roses and last year she sold 326. How many did she sell altogether? ________

13 1·6, 1·7, 1·8, ____, ____, ____, ____

20:4 Extension ☐ out of 7

1 6 eggs can fit in my carton. How many eggs could fit if the egg carton was:

a twice as long? ________

b twice as long and twice as wide? ________

2 Draw all the lines of symmetry on this square. Total = ________

3 Draw all the diagonals on this square. Total = ________

4 **a** 32 + 32 + 32 + 32 ________

b 32 × 3 ________

5 This figure has triangles of different sizes. How many triangles altogether? ________

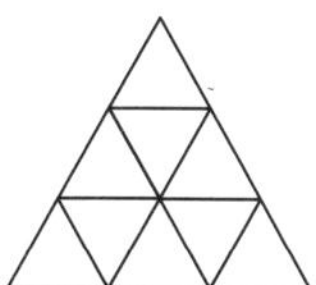

6 How many identical trapeziums could be cut from a regular hexagon? ________

7 One day has 24 hours. How many days in:

a 48 hours? ________ **b** 72 hours? ________

Challenge

Complete each column as quickly as you can. Once finished, check your answers. Record your time.

3 × 2	____	6 × 10	____
5 × 4	____	5 × 5	____
2 × 10	____	8 × 5	____
2 × 5	____	4 × 8	____
3 × 8	____	1 × 10	____
5 × 8	____	9 × 8	____

Time = ____ seconds

- Throw one dice.
- From the start, move forward the number of spaces shown on the dice.

Start		Go 3 ahead.				Go 4 back.
		Go 6 back.	Go 4 ahead.			
Go to start.	Go 1 back.	Go 4 ahead.			Go 3 back.	Finish

What is the least number of throws needed to reach the finish? ________

21:1 out of 20

1. 19 + 8 ______
2. 4 × 6 ______
3. 2 × 8 ______
4. 5 × 10 ______
5. $\begin{array}{r} 461 \\ +373 \\ \hline \end{array}$
6. 45 + 13 − 13 ______
7. Subtract 9 from 27 ______
8. 12 divided by 4 ______
9. 6 times 5 ______
10. $\begin{array}{r} 749 \\ +164 \\ \hline \end{array}$

11. 57 children visited the pool on Saturday and 35 visited on Sunday. How many visited across the weekend altogether? ______
12. Which of the terms *impossible, unlikely, likely* or *certain* describe the chance of you finding a $30 note tomorrow. ______
13. R = Red, B = Blue, Y = Yellow.

    Which colour is the spinner least likely to land on? ______

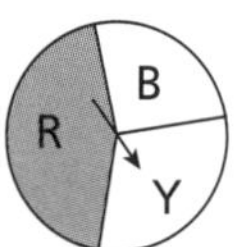

14. The numeral for sixty-eight thousand three hundred and seventy-one ______
15. Round off to the nearest hundred.

    a 427 ______ **b** 857 ______
16. What is the number after 64 728? ______
17. Colour $2\frac{1}{4}$ hexagons.

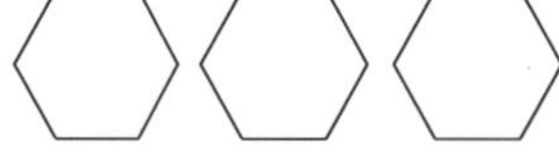

18. What is the month:

    a before August? ______ **b** after March? ______
19. How many cents in $1.25? ______
20. Four more than 35 367. ______

21:2 out of 16

1. 5 × 8 ______
2. 40 ÷ 5 ______
3. 6 × 7 ______
4. 42 ÷ 7 ______
5. $\begin{array}{r} 684 \\ +295 \\ \hline \end{array}$
6. 6 groups of 3 ______
7. 18 shared by 6 ______
8. 23 + ______ = 30
9. 53 + ______ = 60
10. $\begin{array}{r} 582 \\ +259 \\ \hline \end{array}$
11. There are 365 days in a year. How many days in 2 years? ______
12.

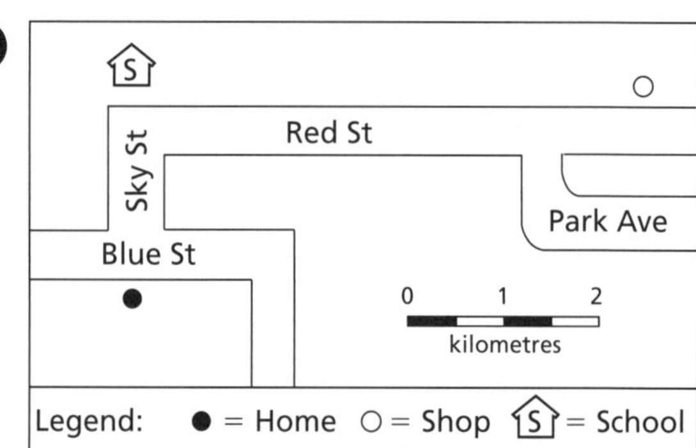

Use the scale to find the distance from home to:

a school ______ **b** the shop ______

c On which street is the shop? ______

d What special feature is 5 km from the school? ______

13. If I toss a 20c coin, is it more likely to land on a head than a tail? ______
14. What day is 3 days after Sunday? ______
15. Write the number that has 4 tens, 6 units, 6 tenths and 7 hundredths. ______
16. The time 15 minutes after 4:35 is ______.

tables

* − 6: 9, 16, 11, 12, 15, 7, 14, 8, 10, 13

* − 7: 9, 13, 10, 17, 11, 16, 12, 15, 7, 14

6 + 7 = 13

13 − 6 = ... 13 + 7 = ...

21:3 ☐ out of 9

1.

hund.	tens	ones
	3	4
+ 1	6	6

2.

hund.	tens	ones
	4	7
+ 5	2	1

3. I bought a jacket for $137 and shoes for $180. How much did I spend? ______

4. Kyle had 25 toy cars and Tyler had twice as many.

a How many toy cars does Tyler have? ______

b How many do they have altogether? ______

5. How would you describe the chance of tossing a head or tail on a dice?

______

6.

A 

B 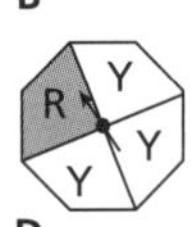

C 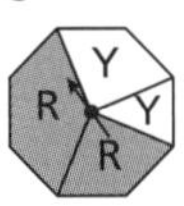

D 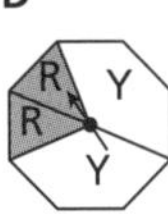

a Which spinner is more likely to land on R than Y? ______

b Which spinner is as likely to land on R as Y? ______

7. How many days in 6 weeks? ______

8. In which month is Christmas? ______

9. What is the abbreviation for:

a centimetres? ______ b kilograms? ______

c litres? ______ d minutes? ______

21:4 ☐ out of 6

Extension

1. 9 triangles have been used to make 3 rows of this pattern. How many triangles are needed to make:

a 4 rows? ______

b 6 rows? ______

2. a Days in a year ______

b Days in a leap year ______

c Days in 4 consecutive years ______

3. This figure has squares of different sizes.

How many squares altogether? ______

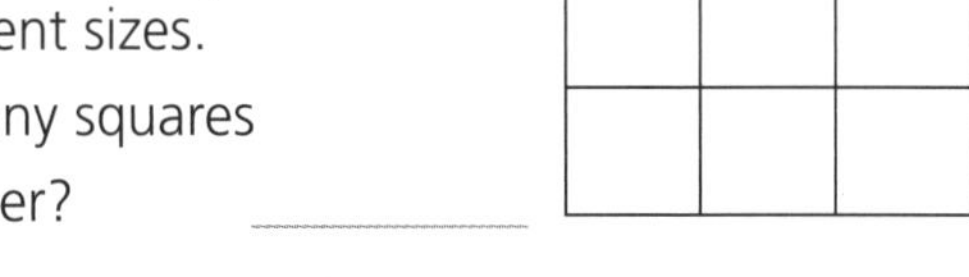

4. How many balls could I buy with $8 if each cost 25 cents? ______

5. How many place-value tens would be needed to reach two metres? ______

6. a 141, 147, 153, ______, ______, ______, ______

b 275, 267, 259, ______, ______, ______, ______

c 365, 377, 389, ______, ______, ______, ______

Challenge

There are 3 children. How many stars could you have so that each child gets a fair share? (Write at least 3 different-sized groups.)

22:1 ☐ out of 14

1. 2 × 7 ______
2. 4 × 7 ______
3. 28 ÷ 4 ______
4. 35 + 24 ______
5. $\begin{array}{r} 594 \\ +258 \\ \hline \end{array}$
6. 10 divided by 5 ______
7. 8 groups of 8 ______
8. 8 shared by 4 ______
9. 40c + 30c ______
10. $\begin{array}{r} 787 \\ +254 \\ \hline \end{array}$
11. Write the number that has 4 tens, 2 units, 3 tenths and 6 hundredths. ______
12.

April						
Sun	Mon	Tue	Wed	Thu	Fri	Sat
	1	2	3	4	5	6
7	8	9	10	11	12	13
14	15	16	17	18	19	20
21	22	23	24	25	26	27
28	29	30				

   a What day is the 19th? ______
   b How many Wednesdays in the month? ______
   c What is the first day of the month? ______
   d What is the day three days before the 26th? ______
13. a 3, 6, 9, ______, ______, ______, ______, ______
   b 45, 38, 31, ______, ______, ______, ______
   c 4, 8, 12, ______, ______, ______, ______, ______
14. Draw the top view of each shape.
   a ☐
   b ☐

22:2 ☐ out of 17

1. 5 × 9 ______
2. 10 × 9 ______
3. 6 × 6 ______
4. 7 × 6 ______
5. $\begin{array}{r} 768 \\ +164 \\ \hline \end{array}$
6. 3 m + 3 m + 3 m ______
7. 1 m − 30 cm ______
8. 3·8, 3·9, ______, ______
9. 32 divided by 4 ______
10. $\begin{array}{r} 693 \\ +368 \\ \hline \end{array}$
11. A dice is rolled. Lachlan wins if an odd number is rolled. Josh wins if a 2 or 4 are rolled. Sarah wins if a 6 is rolled. Who is:
   a most likely to win? ______
   b least likely to win? ______
12. a 45, 50, 55, ______, ______, ______, ______
   b 34, 45, 56, ______, ______, ______, ______
13. What is the day 9 days after Thursday? ______
14. Write the number that has 5 tens, 7 units, 3 tenths and 9 hundredths. ______
15. An edge is a straight side. For each object, how many:

	surfaces?	corners?	edges?
cylinder			
cube			
cone			

16. A ribbon is 1 metre long. If I cut off 30 cm, how much is left? ______
17.

	3	7	4	5	2	10	6	9
× 9								

Concept

In each case, finish the given pattern.

a Add 4.

6					

b Multiply by 2.

2					

c Subtract 9.

90					

d Add 11.

11					

e Add 5.

4					

f Subtract 8.

90					

I am making a 'subtract 5' pattern.

22 − 5

37	32	27	22		

22:3 ☐ out of 9

1

tens	ones
6	1
− 4	3

2

hund.	tens	ones
2	8	1
+ 3	3	4

3 A

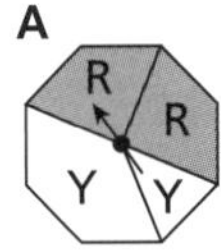

B

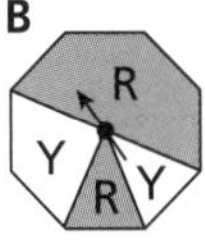

a Which spinner is more likely to stop on red (R)? ______

b Which spinner has an equal chance of stopping on red (R) and yellow (Y)? ______

4 Write the numeral shown.

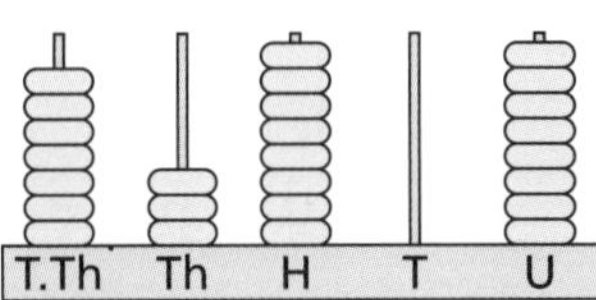

______

5 Colour $2\frac{1}{2}$ of these pentagons.

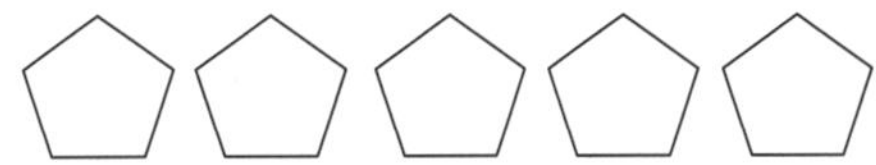

6 There are 11 children in each row. How many children would be in 5 rows? ______

7 Start with a number above 30 and write a pattern with the rule subtract 6.

______, ______, ______, ______, ______, ______

8 On this object, how many:

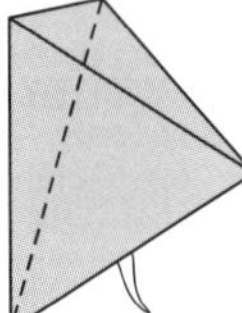

a faces? ______

b edges? ______

9 The value of the 7 in 70 392. ______

Extension 22:4 ☐ out of 5

1 Write a pattern with the rule:

a multiply by 2.

______, ______, ______, ______, ______, ______

b subtract 24.

______, ______, ______, ______, ______, ______

2 What is the smallest number of matchsticks needed to make:

a 2 squares? ______

b 4 squares? ______

c 6 squares? ______

d 8 squares? ______

e 10 squares? ______

3 How many days in 6 weeks? ______

4 Amy earned $15 more than Libby. Lee and Libby earned $32 each.

a How much did Amy earn? ______

b How much did they earn altogether? ______

5 a 12 + 12 + 12 + 12 ______

b 24 + 24 + 24 + 24 ______

Challenge

Make a number pattern using the rule:

a Add 25. ______

b Subtract 30. ______

c Add 9. ______

d Subtract 6. ______

e Add 50. ______

Turn to ID card B on page 7.

Give the answers for these numbers.

(21) ______ sixths shaded

(22) ______ hundredths shaded

(23) ______ tenths shaded

(24) 0· ______ shaded

(25) ______ hundredths covered

(26) decimal ______

(27) ______

$\frac{4}{6}, \frac{5}{6}, \frac{6}{6}, \ldots$

0·43

 • *AUSTRALIAN SIGNPOST MATHS 4 MENTALS* • ISBN 978 0 6557 0884 1

23:1 out of 17

1. 2×7 ____
2. 4×7 ____
3. $21 + 7$ ____
4. $42 + 7$ ____
5. $\begin{array}{r} 681 \\ +132 \\ \hline \end{array}$
6. Multiply 0 by 7. ____
7. 3 groups of 7 ____
8. 14 divided by 2 ____
9. Double 8. ____
10. $\begin{array}{r} 463 \\ +154 \\ \hline \end{array}$

11. Write a pattern with the rule add 3.

____, ____, ____, ____, ____, ____

12. Draw the front and side view of the cube.

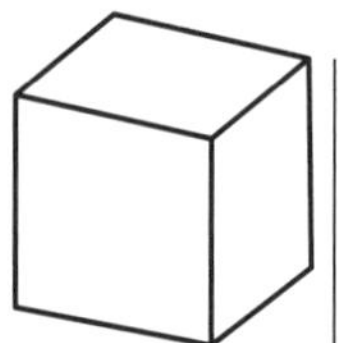

front side

13. A cube has ____ faces, ____ edges and ____ corners.

14. 12 eggs = one dozen
How many eggs in half a dozen? ____

15. Is 3×7 the same as 2×7 plus 7 more? ____

16. Does 7×2 equal 2×7? ____

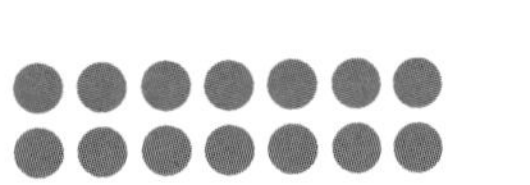

17. **a** Is this maths book heavier than 1 kg? ____
b Is 2L of milk heavier than 1 kg? ____

23:2 out of 17

1. 5×7 ____
2. 10×7 ____
3. 3×7 ____
4. 6×7 ____
5. $\begin{array}{r} 568 \\ +453 \\ \hline \end{array}$
6. 21 divided by 3 ____
7. 7 groups of 7 ____
8. 8 times 7 ____
9. 7, 14, 21, ____, ____
10. $\begin{array}{r} 759 \\ +136 \\ \hline \end{array}$

11. Write a pattern with the rule add 100.

____, ____, ____, ____, ____, ____

12. A  B corny flakes C 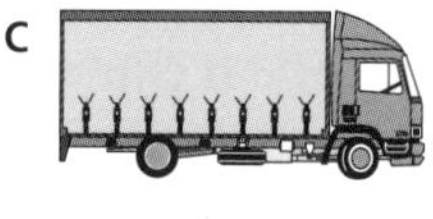

Which of **A**, **B** or **C** would be weighed in:

a kilograms? ____ **b** grams? ____

13. $1\frac{1}{2}$, 2, $2\frac{1}{2}$, ____, ____, ____, ____

14. On this triangular prism, how many:

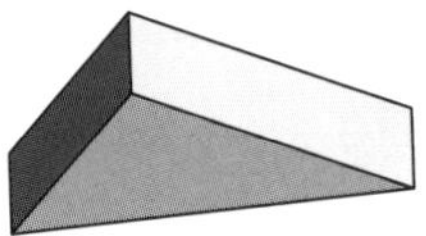

a faces? ____
b edges? ____

15. Is 9×7 the same as 10×7 minus 7? ____

16. Write the short form for:
a 30 grams ____ **b** 20 kilograms ____

17. If today is Tuesday, what day will it be in 10 weeks? ____

× tables

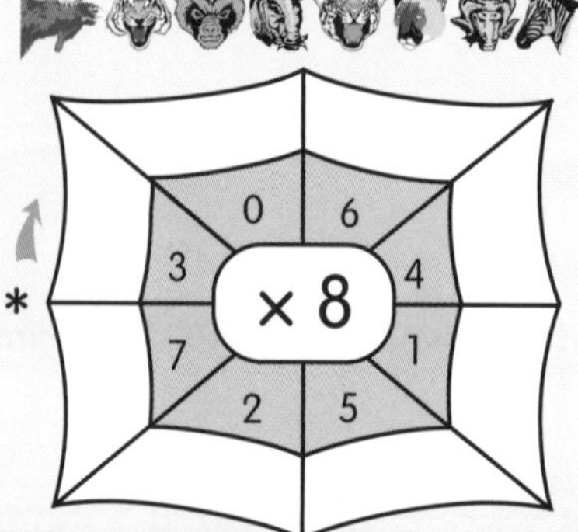

23:3 out of 9

1

hund.	tens	ones
4	6	5
+	4	8

2

hund.	tens	ones
2	9	3
+ 4	1	9

3 How many days in:

a 6 weeks? ______ b 8 weeks? ______

c 10 weeks? ______ d 4 weeks? ______

4 Circle the correctly written measure.

a 9 kg b 9 KG c 9 Kg d 9 Kgs

5 Which is measured in kg? ______

A 

B 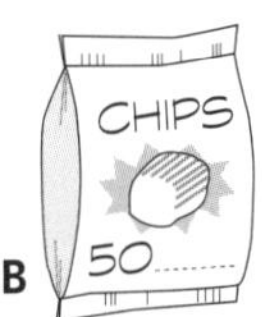

C 

6 Circle each equation that is equal to 12.

3 × 4	4 × 4	8 + 5	6 × 2	20 − 7	5 × 2
3 × 3	2 × 6	14 − 2	4 × 3	6 + 6	12 × 1

7 In a cupboard there were 18 cups, 10 bowls, 16 saucers and 13 plates. How many more cups were there than plates? ______

8 a 637, 643, 649, ______, ______, ______

b 732, 723, 714, ______, ______, ______

9

Colour $3\frac{1}{2}$ of this group.

23:4 out of 8

Extension

1 70 − 11 − 11 − 11 − 11 ______

2 It takes 12 minutes to colour in one page. How long will it take to colour 5 pages? ______

3 a The number of vertices on 8 cubes. ______

b The number of faces on 5 cubes. ______

c The number of edges on 2 cubes. ______

4 I have three surfaces and I can roll. What shape am I? ______

5 a 43 thousands + 32 thousands ______

b 53 thousands + 38 thousands ______

6 What is the change from $100 if you spend:

a $36? ______ b $58? ______

7 I poured 103 mL out of a 1L milk container. How much is left in the container? ______

8 974, 865, 756, ______, ______, ______, ______

Challenge

Write your own number patterns and the rule for each.

______

______

______

______

______

× tables

×7: 1, 2, 3, 4, 5, 6, 7, 8, 9, 10

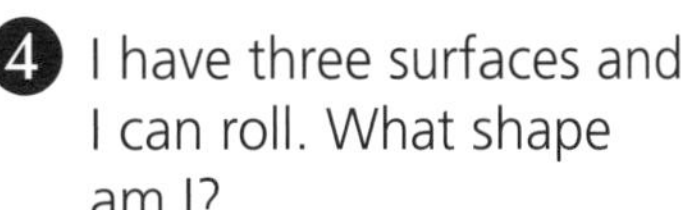

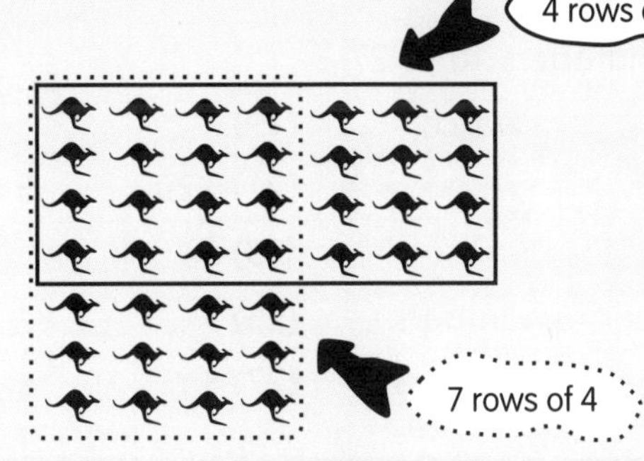

24:1   ___ out of 16

1. 0×7 ____
2. 2×7 ____
3. $14 \div 2$ ____
4. 10×7 ____
5. $\begin{array}{r} 795 \\ -161 \\ \hline \end{array}$
6. 7 shared by 1 ____
7. Triple 7. ____
8. 21 divided by 3 ____
9. 70c − 30c ____
10. $\begin{array}{r} 678 \\ -257 \\ \hline \end{array}$

11. Name the shape below. ____

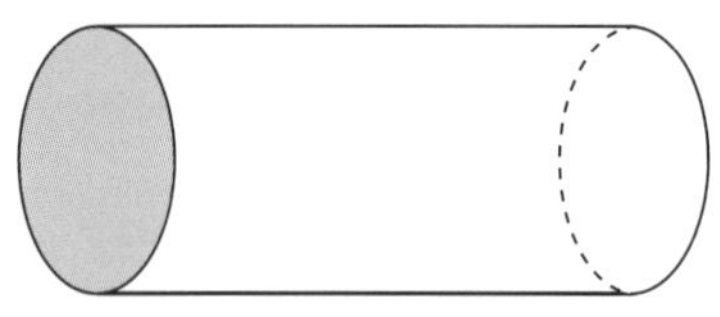

12. On the shape above, how many:
    a surfaces? ____   b corners? ____

13.

Ham	Grapes	Melon
4 kg	2 kg	6 kg

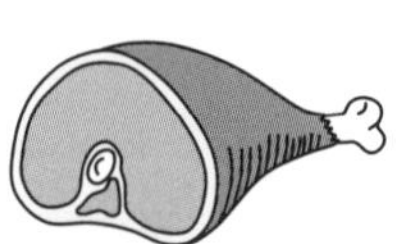
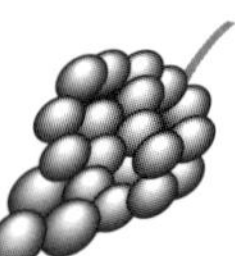

a Put these items in order of mass, from least to most.

____

b The total mass. ____

14. Are 10, 16, 28 and 32 all even numbers? ____
15. How many weeks in 1 year? ____
16. How many days in 1 fortnight? ____

24:2   ___ out of 16

1. 9×7 ____
2. 8×7 ____
3. $37 + 13$ ____
4. $24 + 16$ ____
5. $\begin{array}{r} 624 \\ -203 \\ \hline \end{array}$
6. 30 divided by 3 ____
7. Double 15. ____
8. 5·8, 5·9, ____, ____
9. 28 shared by 7 ____
10. $\begin{array}{r} 576 \\ -316 \\ \hline \end{array}$

11. 

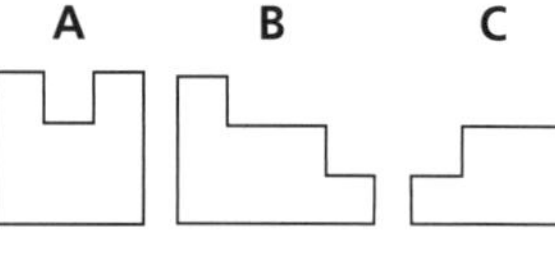

Which shape can the child see? ____

12. a Grams in 1 kg 400 g ____
    b Kilograms in 5000 g ____
    c Grams in $\frac{1}{2}$ kg ____
    d Grams in $5\frac{1}{4}$ kg ____
13. The time is ____ to ____.

14. How many weeks in:
    a 28 days? ____
    b 42 days? ____
15. How many weeks in:
    a 2 years? ____   b 3 years? ____
16. Write the mixed numeral for:
    a $\frac{4}{3}$ ☐   b $\frac{5}{2}$ ☐   c $\frac{12}{10}$ ☐

Turn to ID card A on page 6.
Give the answers for numbers 18 to 27

(18) ____ watch   (19) ____ clock
(20) 5 ____ 7   (21) 20 ____ 5
(22) 18 ____ 6   (23) ____ numbers
(24) ____ numbers   (25) ____ numbers
(26) ____ numbers   (27) ____

24:3 [ ] out of 10

1

H	T	U
8	6	5
− 2	4	1

2

H	T	U
7	9	4
− 5	2	0

3 What shapes have been used in this picture? ______

4 Circle each equation that is equal to 24.

5 × 5	20 + 4	8 × 3	30 − 6	6 × 4	18 + 6
18 + 7	27 − 5	12 × 2	29 − 5	3 × 8	34 − 10
2 × 12	15 + 9	6 × 4	10 + 14	5 × 7	48 ÷ 2

5 True (T) or false (F)?
A brick is heavier than 1 kg. ______

6 Complete the labels.

a

4: [ ]

[ ] past 4

b

10: [ ]

[ ] to 11

7 800 000 + 30 000 + 4000 + 6 ______

8 Seconds in 1 minute. ______

9 7 units + 5 tenths + 8 hundredths = ______

10 **a** 30 minutes after 4:36 pm is ______.
b 30 minutes before 8:42 am is ______.

24:4 [ ] out of 5

Extension

1 The same number must go in each of the three boxes.

Give two possible answers. ______ or ______

2 Draw the different views of this shape.

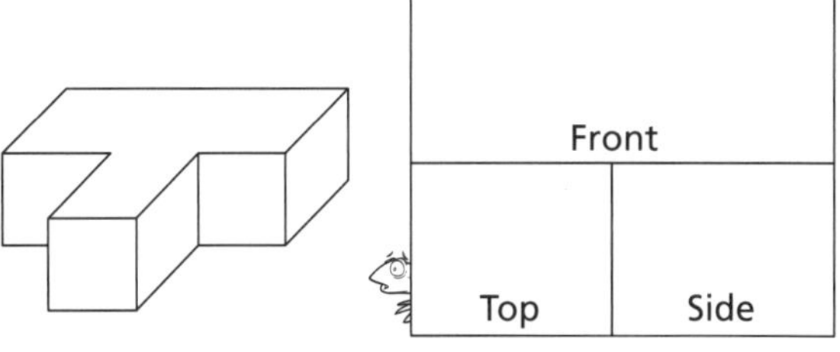

3 In this solid symmetrical shape, how many blocks cannot be seen from this angle? ______

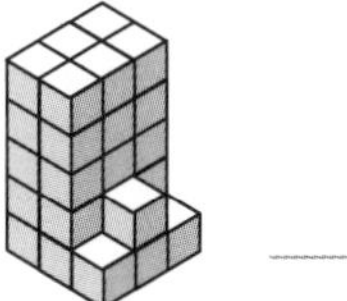

4 **a** 34 minutes after 2:06. ______

b 10 minutes before 2:06. ______

5 If the day after tomorrow will be Friday, what day was it 10 days ago? ______

Challenge

Write what you know about the number 46•84.

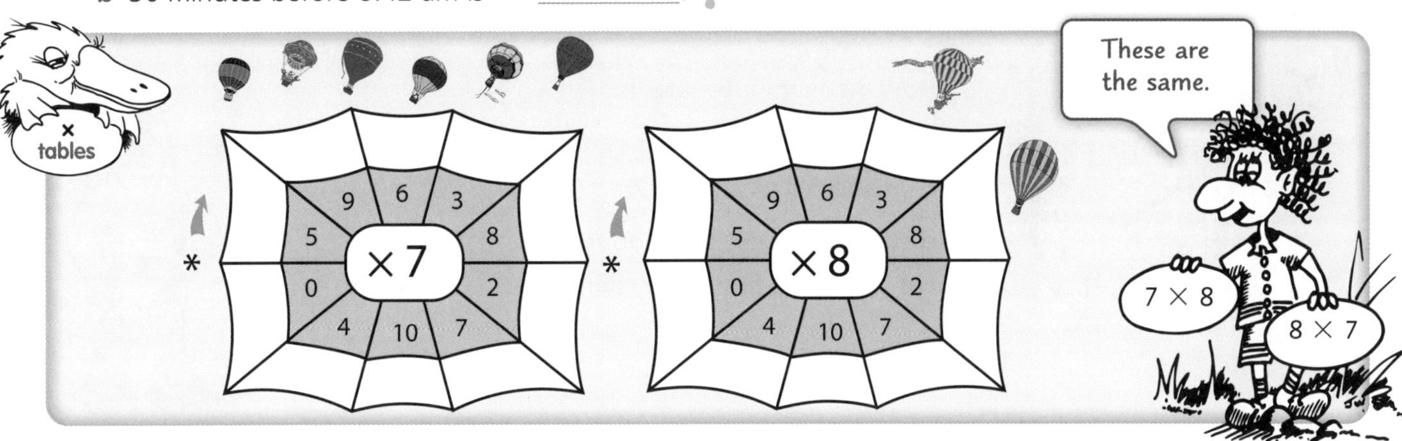

25:1 — out of 17

1. 8 × 2 ____
2. 4 × 3 ____
3. 8 × 3 ____
4. 15 + 5 ____
5. $\begin{array}{r} 395 \\ -287 \\ \hline \end{array}$
6. 20 more than 15 ____
7. 20 less than 54 ____
8. Divide 24 by 4. ____
9. Multiply 6 by 10. ____
10. $\begin{array}{r} 730 \\ -314 \\ \hline \end{array}$
11. Write the improper fraction and mixed numeral for the part coloured.
12. a Write the digital time. ____
    b The time is ____ minutes to ____.
13. How many school days in 4 weeks? ____
14. 100 centimetres = ____ metre
15. Write the short form for:
    a centimetres ____ b metres ____
16. Which unit (m, cm or mm) would you use to measure the width of a:
    a tennis court? ____ b cat? ____
    c pencil? ____ d cup? ____
17. a 24, 28, 32, ____, ____, ____, ____
    b 75, 71, 67, ____, ____, ____, ____

25:2 — out of 18

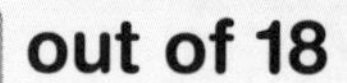

1. 3 × 7 ____
2. 6 × 7 ____
3. 4 × 8 ____
4. 8 × 8 ____
5. $\begin{array}{r} 832 \\ -386 \\ \hline \end{array}$
6. 32 shared by 4 ____
7. 24 divided by 6 ____
8. Multiply 6 by 3. ____
9. Add 5 to double 7. ____
10. $\begin{array}{r} 947 \\ -379 \\ \hline \end{array}$

11. Would watermelons be weighed in kilograms? ____
12. a The digital time is ____ : ____.
    b The analog time is ____ past ____.

13. If it is Tuesday the 15th, what will it be in 9 days time? ____
14. Write 5 m 37 cm as centimetres. ____
15. Write 267 cm as metres and centimetres. ____
16. This line is ____ cm long.
17. From the numbers 1, 4, 8, 30 and 50, choose the best estimate for the length of a car in metres. ____
18. The measure is ____ kg .

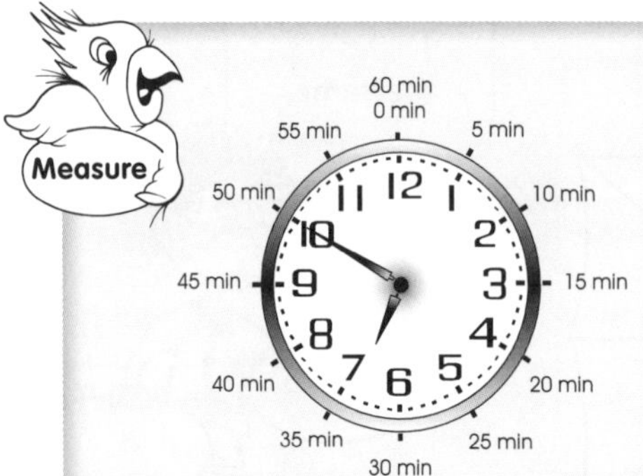

How many minutes are in:

a one hour? ____ b half an hour? ____

c a quarter of an hour? ____

d three quarters of an hour? ____

For this clock, how many minutes:

e is it after 6 o'clock? ____

f is it before 7 o'clock? ____

 ISBN 978 0 6557 0884 1

25:3 ☐ out of 8

1.

	H	T	U
	5	6	1
−	1	3	8

2.

	H	T	U
	6	7	0
−	4	9	8

3. This digital time:

a is read as

______

b means ____ minutes past ____

4. Complete the labels.

a

3 : ☐

☐ to 4

b

6 : ☐

☐ to 7

5. Measure the length of:

a ______ cm ______ mm

b ______ cm ______ mm

6. Estimate, then find the perimeter of this triangle.

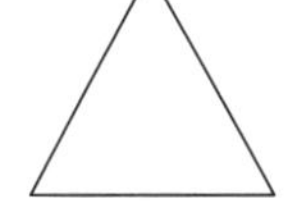

Estimate: ______ mm

Measure: ______ mm

7. a 600 + 300 ____ b 800 + 500 ____

8. Use the jump strategy to find:

58 + 64 ______

58

25:4 ☐ out of 4

Extension

1. How many hours are in 4 days? ______

2. Write an equation in each box that is equal to 32.

3. I live 80 m from school.
How far did I walk to school and back in one school week? ______

4.

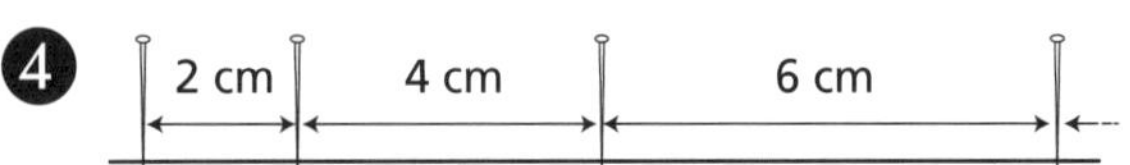

Pins were stuck into the floor using the pattern shown above.

What was the distance between:

a the 5th and 6th pins? ______

b the 1st and 6th pins? ______

Challenge

Complete these columns of facts as quickly as you can. Once complete, check your answers and record your time.

2 × 7	______	8 × 2	______
6 × 5	______	3 × 7	______
5 × 8	______	10 × 7	______
9 × 7	______	6 × 7	______
4 × 7	______	3 × 8	______
6 × 6	______	8 × 7	______
6 × 8	______	7 × 8	______
7 × 7	______	5 × 7	______

Time = ______ seconds

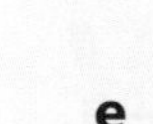

Complete, then learn these facts. (See page 84.)

Half of 60 is 30.

a ______ seconds = 1 minute
b ______ hours = 1 day
c ______ minutes = 1 hour
d ______ days = 1 week
e ______ minutes = half an hour
f 1 fortnight = ______ weeks
g ______ days = one leap year
h ______ days = one ordinary year

26:1

out of 17

1. 156 + 4 ____
2. 175 + 5 ____
3. 35 + 52 ____
4. 14 + 83 ____
5. 593 − 186
6. 23 + 9 ____
7. 23 + 19 ____
8. 12 + 6 + 8 ____
9. 15 + 3 + 5 ____
10. 382 − 196

11. The digital time:
    a is read as ____
    b means ____ to ____

12. Write 2 cm 7 mm as millimetres. ____
13. a Measure each length in millimetres .

____mm

____mm

    b The difference in length is ____.
14. The measure is ____ kg.

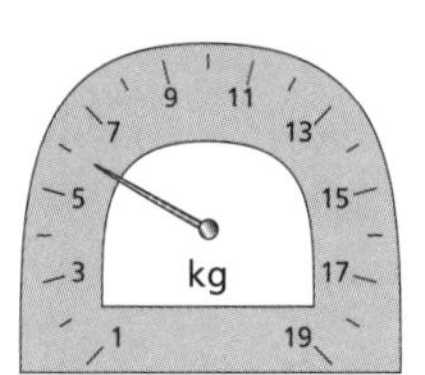

15. 7 tenths + 9 hundredths = ____
16. Bridge to 10 to find:
    a 37 + 4 ____ b 68 + 5 ____
    c 159 + 6 ____ d 245+ 7 ____
17. Use the split strategy to find:
    a 346 + 142 ____ b 624 + 152 ____

26:2

out of 17

1. 5 + 7 ____
2. 65 + 7 ____
3. 365 + 7 ____
4. 425 + 19 ____
5. 653 − 396
6. 417 + 50 − 3 ____
7. 456 + 7 ____
8. 346 + 213 ____
9. 365 + 100 − 5 ____
10. 782 − 597

11. How many minutes is it after two o'clock? ____

12. 36 months = ____ years.
13. 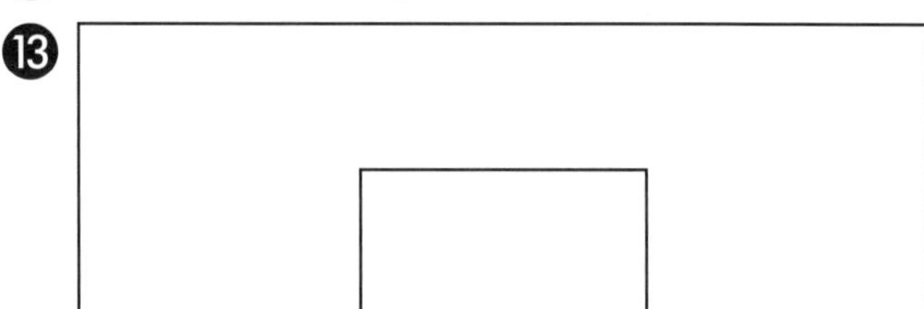
    a What is the perimeter of this shape? ____
    b How many place-value ones blocks would be needed to cover this area? ____
14. a Estimate your height. ____
    b Is the height of a door about 2 m? ____
15. Use the short form to write:
    a 13 square metres ____
    b 34 square centimetres ____
16. Round off then compensate to find:
    a 135 + 19 ____ b 264 + 29 ____
    c 369 + 32 ____ d 138 + 43 ____
17. a 900 + 400 ____ b 700 + 700 ____

Turn to ID card B on page 7.
Give the answers for these numbers.

(1) ____ line (2) ____ line
(3) ____ line (4) ____ lines
(5) ____ line (6) ____ of ____
(19) ____ expander (20) ____

26:3

out of 9

1

H	T	U
7	1	4
− 3	4	6

2

H	T	U
5	2	0
− 1	7	2

3 What is the digital time?

a 

b 

________ ________

4 4 weeks = ____ days

5 Estimate then measure the length of this bar.

Estimate = ______ cm Measure = ______ cm

Measure = ______ mm

6 Write 5 m 21 cm as centimetres. ______

7 Would you use m^2 or cm^2 to measure the area of:

a the classroom floor? ______

b a mobile phone? ______

8 118 + 27 ______

118

9 Round off then compensate to find:

a 352 + 19 ______ b 637 + 39 ______

c 618 + 43 ______ d 138 + 434 ______

26:4

Extension

out of 6

1 How many hours in 1 week? ______

2 Use mental strategies to find:

a 197 + 35 + 33 + 15 + 19 ______

b 463 + 97 ______

c 364 + 298 ______

d 177 + 624 ______

e 647 − 55 ______

f 476 − 97 ______

3

How many different pairs could be chosen from these people? ______

4 How many 20c coins have the same value as:

a \$7.40? ______ b \$20? ______

5 How many sides on 12 hexagons? ______

6 a

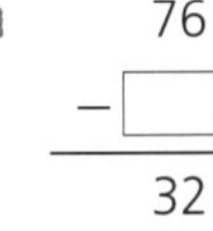

76
− ☐
32

b

42
− ☐
17

Challenge

Use scales to find and measure objects. List the object and the weight for each below.

______

______

______

______

Roman numerals

Use this clock to give the Roman numeral for:

a 4 ____ b 5 ____ c 9 ____ d 10 ____

e 11 ____ f 12 ____ g 2 ____ h 8 ____

Write our numeral for:

i IX ____ j XI ____ k VI ____ l IV ____

m III ____ n VIII ____ o XII ____ p VII ____

27:1 [ ] out of 17

1. 38 + 9 ______
2. 38 + 19 ______
3. 38 + 29 ______
4. 30 − 6 ______
5.
```
  369
− 147
```
6. 15 + 5 + 7 ______
7. 12 + ______ = 20
8. 18 + 7 ______
9. 23 − 8 ______
10.
```
  530
− 373
```

11. Would you measure the area of a wall in m^2 or cm^2? ______
12. 88 + 34 ______

88 ⟶

13. Build to 10 to find:
 a 56 + 9 ______ b 38 + 5 ______
 c 168 + 7 ______ d 147 + 9 ______
14. Complete:

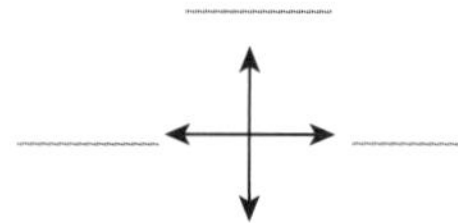

15. 卌 卌 卌 卌 卌 III
 This tally stands for ______.
16. 3 × 5 = ______
 15 ÷ 3 = ______
 ______ × 5 = 15

17. a How many toes on 7 people? ______
 b How many legs on 7 octopuses? ______
 c How many sides on 7 hexagons? ______

27:2 [ ] out of 18

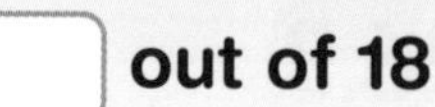

1. 637 + 8 ______
2. 359 + 6 ______
3. 265 + 134 ______
4. 300 + 200 ______
5.
```
  600
− 361
```
6. 900 minus 200 ______
7. 377 add 96 ______
8. 236 plus 153 ______
9. Does $\frac{5}{10}$ = 0·5? ______
10.
```
  50
− 29
```

11. Use the short form to write:
 a 64 square metres ______
 b 92 square centimetres ______
12. 139 + 42 ______

139 ⟶

13. a 31 + 31 + 31 + 31 ______
 b 29 + 29 + 29 + 29 ______
14. If 7 + 5 = 12, find:
 a 27 + 5 ______ b 67 + 5 ______
 c 147 + 5 ______ d 257+ 5 ______
15. My friend and I are back to back. She is facing east. Which direction am I facing? ______
16. 4 × 3 = ______ 3 × 4 = ______
 12 ÷ 3 = ______ 12 ÷ 4 = ______
17. There are 5 monkeys in each cage. How many monkeys are in 6 cages? ______
18. a A square has ______ sides and ______ angles.

Complete, then learn these facts. (See page 84.)

a 1 year = ______ weeks
b 1 leap year = ______ days
c 1 year = ______ months
d 1 decade = ______ years
e 1 year = ______ days
f 1 century = ______ years
g 1 day = ______ hours

27:3 ☐ out of 8

1. H T U
   1 9 9 + 1
   ~~2 0 0~~
   − 5 7
   ______
   ______

2. H T U
   6 9 9 + 1
   ~~7 0 0~~
   − 2 4
   ______
   ______

3. Look for compatible numbers to find:
   a 746 + 153 ______ b 342 + 154 ______
4. Round off then compensate to find:
   a 538 + 48 ______ b 428 + 97 ______
5. Use the split strategy to find:
   a 382 + 417 ______ b 637 + 252 ______
6. What direction is:
   a Pai from A? ______
   b Mun from A? ______

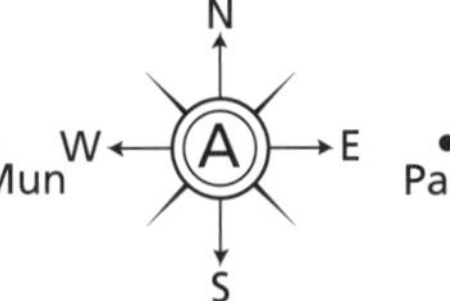

7.

Goals scored	
Tim ~~||||~~ ~~||||~~ |	Cailin ~~||||~~ ||
Kylie ~~||||~~ ||||	Emmy |||

   a How many more goals did Kylie score than Emmy? ______
   b Who scored 7 goals? ______
   c How many goals were scored? ______
   d How many did Kylie score? ______
8. 185 + 36

   185 ————→

27:4 Extension ☐ out of 9

1. 1, 10, 100, ______, ______, ______
2. How many shoes are in 116 pairs? ______
3. How many 60c stamps could I buy for $5? ______
4. How many cards could I buy with $8 if each card cost 25c? ______
5. Choose a mental strategy to find:
   a 849 + 356 ______ b 584 − 97 ______
   c 586 + 35 ______ d 639 + 48 ______
6. 4 books of the same size have 160 pages altogether. How many pages has each book? ______
7. How many 20c coins have the same value as:
   a $3.40? ______ b $12? ______
8. a What is our numeral for XXIV? ______
   b What is our numeral for XXXVII? ______
9. Decrease $20 by $7 and then double the result. ______

Challenge

What facts are shown in this data display?

Name	Books read
Deon	~~||||~~ ~~||||~~ ~~||||~~ |
Jean	~~||||~~ ~~||||~~ ||
Aimee	~~||||~~ |
Lisa	~~||||~~ ~~||||~~

______
______
______
______
______
______
______

Roman numerals

1	I	one finger
5	V	one hand
10	X	two V's
50	L	half of a C
100	C	centum = 100

Write our numeral for:

a V ______ **b** XV ______ **c** LX ______
d CLI ______ **e** CLXV ______ **f** CLXVI ______

Write the Roman numeral for:

g 25 ______ **h** 110 ______ **i** 221 ______
j 207 ______ **k** 67 ______ **l** 333 ______

28:1 ☐ out of 17

1. 8 + 7 ______
2. 28 + 7 ______
3. 148 + 7 ______
4. 65 − 6 ______
5. $\begin{array}{r} 150 \\ -\ 32 \\ \hline \end{array}$
6. Multiply 3 by 5. ______
7. 6 times 5 ______
8. Half of 28 ______
9. 18 shared by 3 ______
10. $\begin{array}{r} 300 \\ -257 \\ \hline \end{array}$

11. Complete:

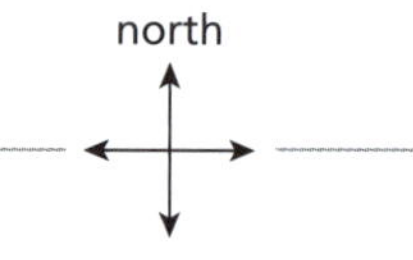

12. What number is represented by this tally? ______

𝍸 𝍸 𝍸 𝍸 |||

13. ÷ This is the d ______ sign.

14. 2 × 3 = ______

6 ÷ 2 = ______

15. If 8 cakes were shared equally among four people, how many would each person get? ______

16.

    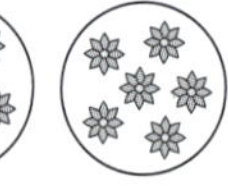

a 5 × 5 = ______ b 25 ÷ 5 = ______

c 3 × 5 = ______ d 15 ÷ 5 = ______

17. 45 637, 45 638, 45 639, ______, ______, ______, ______

28:2 ☐ out of 15

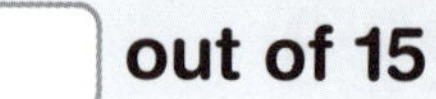

1. 537 + 98 ______
2. 259 + 97 ______
3. 358 + 9 ______
4. 325 + 8 ______
5. $\begin{array}{r} 500 \\ -347 \\ \hline \end{array}$
6. Add 500 and 450. ______
7. 300 subtract 28 ______
8. 187 + 45 + 13 + 5 ______
9. 152 minus 27. ______
10. $\begin{array}{r} 800 \\ -637 \\ \hline \end{array}$

11. 197 + 246 ______

(number line starting at 197)

12. My friend and I are back to back.
He is facing north.
Which direction am I facing? ______

13.

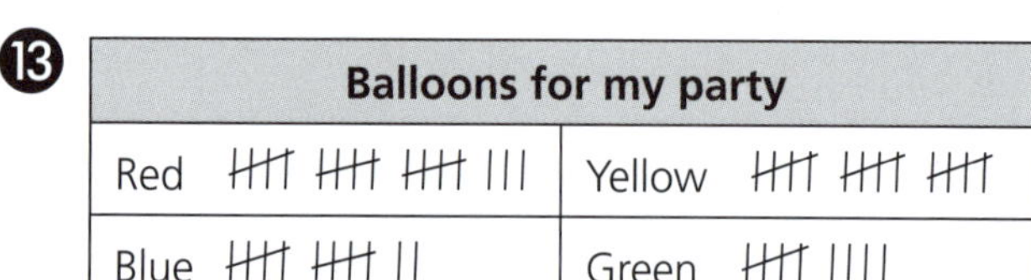

Balloons for my party	
Red 𝍸 𝍸 𝍸 |||	Yellow 𝍸 𝍸 𝍸
Blue 𝍸 𝍸 ||	Green 𝍸 ||||

a How many balloons in total? ______

b How many more red balloons are there than yellow? ______

c How many red and green balloons? ______

14. Share 9 lollies among 3 girls.

One share = ______ 9 ÷ 3 = ______

15. Measure and compare the length of objects in your pencil case. Write what you found.

______

Division using the number line

How many 4s in 12?

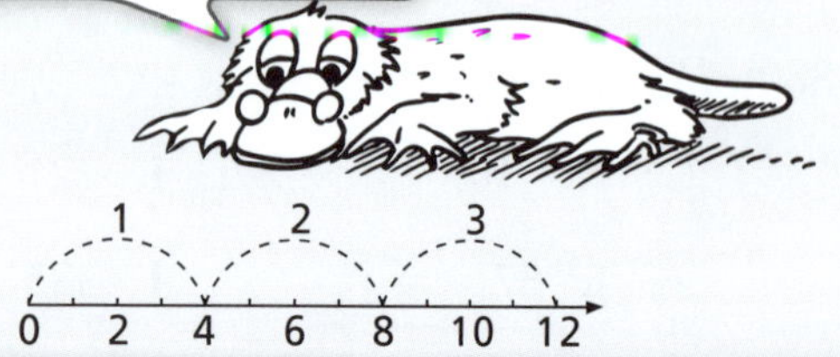

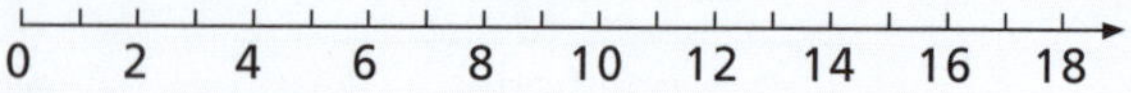

0 2 4 6 8 10 12 14 16 18

Use this number line to find how many:

a 5s in 15 ______ b 3s in 15 ______

c 6s in 18 ______ d 3s in 18 ______

e 2s in 14 ______ f 4s in 16 ______

28:3

out of 7

1

H	T	U	
8	9	9	+ 1
~~9~~	~~0~~	~~0~~	
−	4	6	
			+ 1

= ______

2

H	T	U	
4	9	9	+ 1
~~5~~	~~0~~	~~0~~	
−	6	9	
			+ 1

= ______

3 How many 2s in twenty? ______

4 **a** 3 rows of 6 = ______

b 6 x 3 = ______

c 3 rows of 8 = ______

d 8 x 3 = ______

5 **a** Use the picture above to find how many groups of 3 are in 24. ______

b 24 ÷ 3 = ______ **c** 24 ÷ 8 = ______

6 What is in position:

a 2C? ______

b 3B? ______

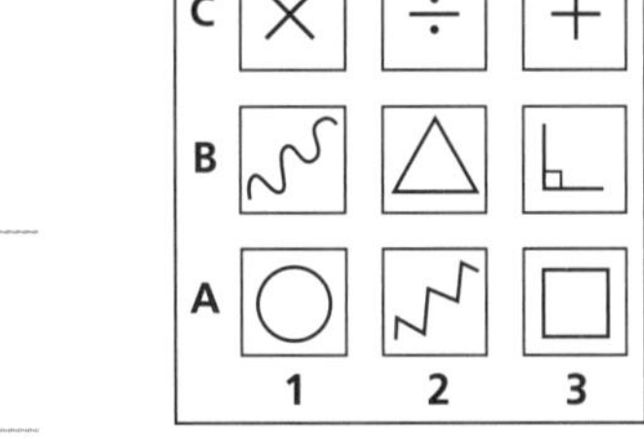

From the grid above, give the position of the:

c triangle. ______

d wavy line. ______

7

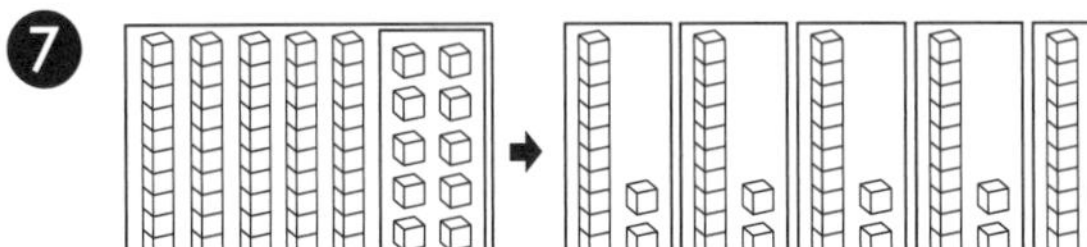

How many 12s are in 60? ______

28:4

Extension

out of 10

1 A desk weighs the same as 12 chairs. How many chairs weigh the same as 3 desks? ______

2 Yuan drew 3 trees. On each he drew 26 branches. How many branches did he draw altogether? ______

3 Number of wheels on 9 trucks. ______

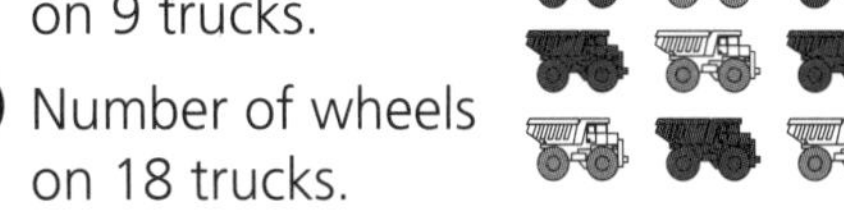

4 Number of wheels on 18 trucks. ______

5 Naomi walked 15 km east and then 23 km west. How far was she then from her starting point? ______

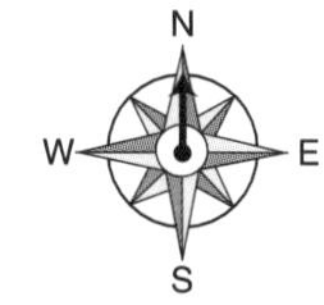

6 How many pairs in 18 shoes? ______

7 If 13 + 13 + 13 + 13 + 13 + 13 = 78, then how many 13s in 78? ______

8 I need $12 to buy one tie. How many ties can I buy with $77? ______

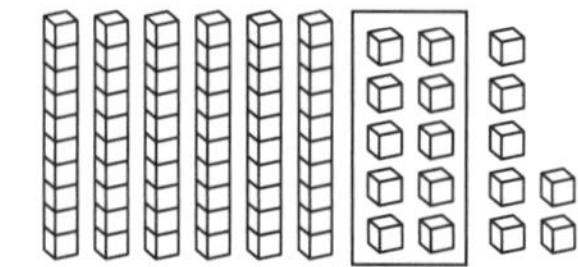

9 Share 55 stickers among 5 children.
One share = ______, 55 ÷ 5 = ______

10 944 take away 356. ______

Challenge

Do you think the letter 'p' or 'b' has been used most on this page? ______
Use tally marks to record how many times each letter has been written.

		Total
p		
b		

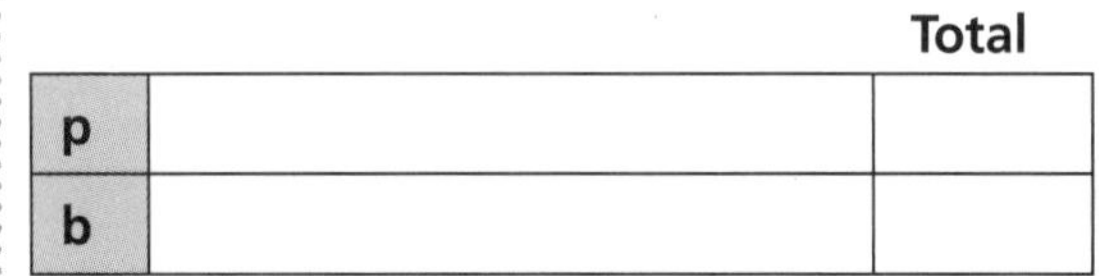

29:1 ☐ out of 14

1. 64 + 19 ______
2. 33 + 29 ______
3. 86 − 21 ______
4. 71 − 52 ______
5. $\begin{array}{r} 500 \\ -323 \\ \hline \end{array}$
6. 123 more than 59 ______
7. 721 minus 63 ______
8. 842 take away 59 ______
9. 149 + 453 ______
10. $\begin{array}{r} 450 \\ -243 \\ \hline \end{array}$
11. ____ × 4 = 16 so 16 ÷ 4 = ____
12. 3 × 4 = ______

    12 ÷ 3 = ______

13.

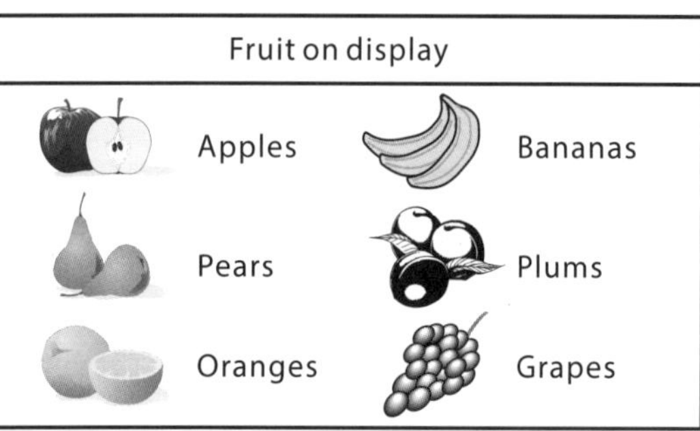

Fruit on display	
Apples	Bananas
Pears	Plums
Oranges	Grapes

   a Which fruit is on the bottom right? ______
   b Which fruit is opposite the pears? ______
   c Which fruit is above the plums? ______
   d How many of the names of these fruits start with 'p'? ______
   e What fraction of the names starts with 'p'? ___ out of ______
14. 450 000 + 6000 + 900 + 20 + 1 ______

______

29:2  ☐ out of 16 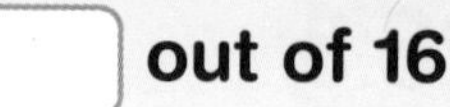

1. 69 + 34 ______
2. 75 + 25 ______
3. 100 − 45 ______
4. 97 − 48 ______
5. $\begin{array}{r} 800 \\ -325 \\ \hline \end{array}$
6. 56 less than 162 ______
7. 253 subtract 60 ______
8. 467 plus 99 ______
9. Does $\frac{6}{10} = 0{\cdot}6$? ______
10. $\begin{array}{r} 400 \\ -256 \\ \hline \end{array}$
11. Circle the mixed number. $\frac{9}{2}$, $\frac{6}{8}$, $7\frac{1}{2}$
12. Share 12 apples among 3 boys.

    One share = ______ 12 ÷ 3 = ______
13. Lia bought 4 mugs for $20. How much did one cost?

    ______
14. 50 cards are shared equally. If each player has 5 cards, how many players are there?

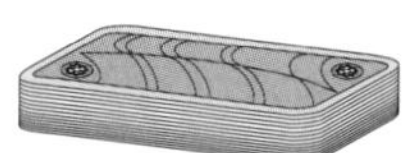

15. Give the coordinates of:

    a W ______
    b X ______
    c Y ______
    d Z ______

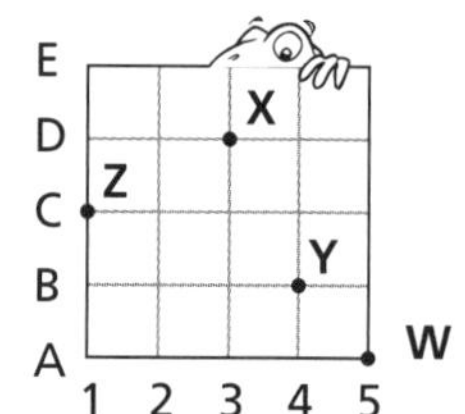

16. Find and measure the length of 6 pencils in mm. Write your results. Tell someone what you can learn from this.

______

______

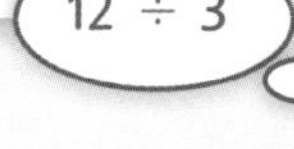

Concept

The division symbol, ÷

12 ÷ 3 means:

How many groups of 3 in 12?
OR **What is one share, if 12 is shared among 3?**

Therefore: **12 ÷ 3 = 4**

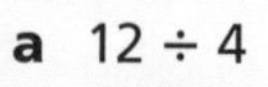

a 12 ÷ 4 ______ **b** 12 ÷ 2 ______
c 12 ÷ 6 ______ **d** 12 ÷ 3 ______
e 20 ÷ 2 ______ **f** 20 ÷ 10 ______
g 15 ÷ 3 ______ **h** 15 ÷ 5 ______
i 28 ÷ 4 ______ **j** 70 ÷ 10 ______

29:3 out of 14

1. H T U
   8 9 9 + 1 → 9 0 0
   − 6 7
   ______ + 1
   = ______

2. H T U
   4 9 9 + 1 → 5 0 0
   − 9 8
   ______ + 1
   = ______

3. 1)9
4. 5)15
5. 4)8
6. 20 ÷ 5 = ______ or ______ × 5 = 20
7. Sharing means d______.
8. a Share 16 crayons between 2 students.

   b One share = ______
   c 16 ÷ 2 = ______
9. 2 × 8 = ______ so 8 × 2 = ______
10. 450 000 + 8000 + 200 + 90 + 7 ______
11. 3 × 6 = ______
    18 ÷ 3 = ______
12. How long does it take to travel from:
    a Blacktown to Pendle Hill? ______
    b Seven Hills to Parramatta? ______
    c Toongabbie to Westmead? ______

Train Timetable	
Blacktown	8:51
Seven Hills	8:54
Toongabbie	8:57
Pendle Hill	8:59
Wentworthville	9:01
Westmead	9:03
Parramatta	9:06

13. 45 people booked a trip on a bus. If the bus left with only 27 people on it, how many missed the bus? ______
14. 540 000 + 7000 + 200 + 80 + 5 ______

29:4 Extension out of 9

1. Share 56 dolls among 4 shops. Each shop is given ______.
2. Of the 39 houses in a street, how many have even numbers? ______
3. If one pizza is needed to feed 5 people, how many are needed to feed 45 people? ______
4. If 17 × 18 = 306, what does 18 × 18 equal? ______
5. If 2 × □ = 18, 3 × △ = 24 and 6 × ◇ = 36, find □ + △ − ◇. ______
6. 197 + 46 ______
   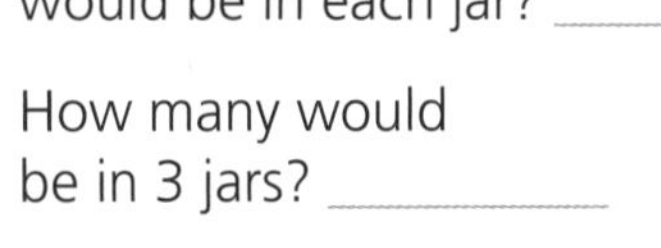

7. There were 224 lollies in 4 jars. How many lollies would be in each jar? ______
   How many would be in 3 jars? ______

8. 600 + 23 000 + 9 + 10 + 400 000 ______
9. 24 more than 45 213 . ______

Challenge

Make up a number story for 20 ÷ 5, then find the answer.

______

N E W S

This word is made from North, East, West, South

N
W E
S

The needle of a compass always points north.

Which parts of Australia are in the:

a west? ______
b north? ______
c east? ______
d south? ______

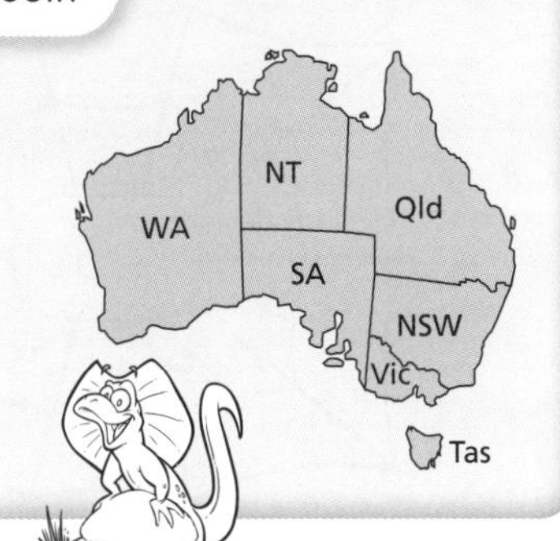

30:1

out of 15

1. $12 \div 2$ ____
2. $14 \div 2$ ____
3. $20 \div 2$ ____
4. $40 \div 4$ ____
5. $\begin{array}{r} 354 \\ +437 \\ \hline \end{array}$
6. 16 shared by 4 ____
7. Divide 25 by 5. ____
8. 30 divided by 3 ____
9. 14 less than 30 ____
10. $\begin{array}{r} 726 \\ -518 \\ \hline \end{array}$
11. $2 \times 4 =$ ____
    $8 \div 2 =$ ____
12. **a** These tops show

    ____ rows of ____ = ____

    b $6 \times 5 =$ ____ $5 \times 6 =$ ____
13. Share 30 tops between 5.
    ____ each $30 \div 5 =$ ____
14. 

    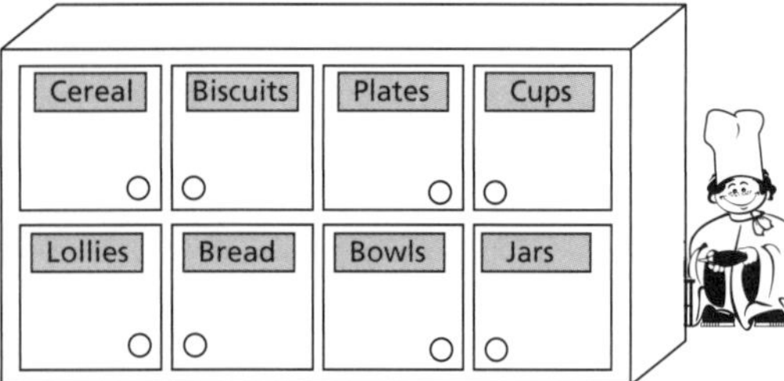

    a What is kept second from the left, on the top row? ____

    b Where is the bread kept? ____

    ____
15. **a** 4 legs.
    How many emus? ____

    b 12 legs.
    How many emus? ____

30:2

out of 16

1. $18 \div 6$ ____
2. $63 \div 9$ ____
3. $64 \div 8$ ____
4. $15 \div 3$ ____
5. $\begin{array}{r} 437 \\ +263 \\ \hline \end{array}$
6. 24 divided by 6 ____
7. 28 shared by 7 ____
8. Does $\frac{1}{10} = 0{\cdot}1$? ____
9. 3·4, 3·5, ____, ____
10. $\begin{array}{r} 572 \\ -284 \\ \hline \end{array}$

11. ____ $\times 6 = 30$ so $30 \div 6 =$ ____
12. We bought 36 oranges and shared them into 4 equal bags.

    How many in are each bag? ____
13. Circle the improper fraction. $\frac{7}{4}$, $\frac{8}{10}$, $5\frac{2}{3}$
14. Circle: An even number multiplied by an odd number equals an **odd** / **even** number.
15. On this 2027 calendar:

DECEMBER						
Sun	Mon	Tue	Wed	Thu	Fri	Sat
		1	2	3	4	5
6	7	8	9	10	11	12
13	14	15	16	17	18	19
20	21	22	23	24	25	26
27	28	29	30	31		

    a What day is December 12? ____

    b How many Thursdays are in December? ____

    c Write the short date for the second Tuesday.

    ____

    d How many days is it from the first Friday to the last Monday of December? ____
16. **a** 30 minutes after 4:25 pm? ____

    b 30 minutes before 6:29 am? ____

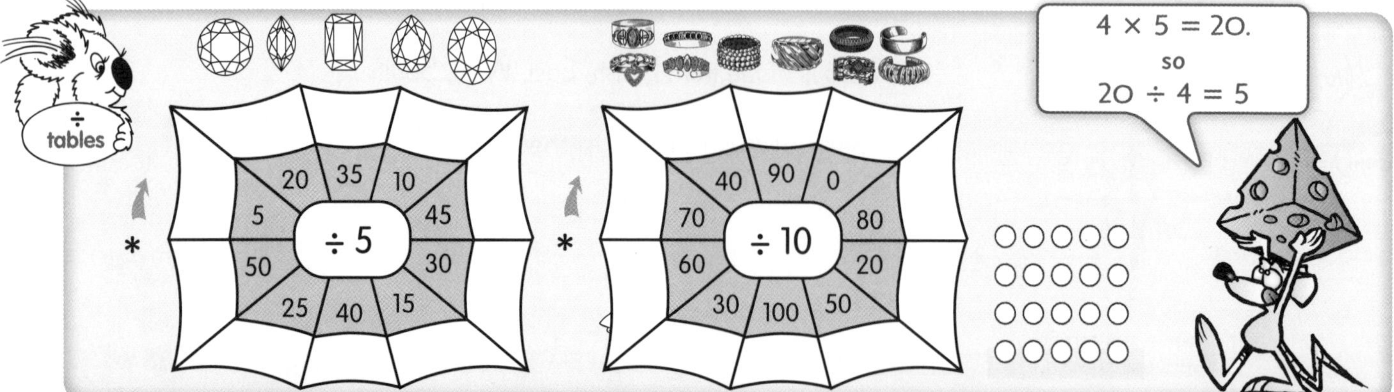

30:3

out of 12

1

hund.	tens	ones
1	4	2
+	5	3

2

hund.	tens	ones
	\$8	7
+	\$5	2

3 5)10 4 6)24 5 8)48

6 3 × 5 = ______

15 ÷ 3 = ______

7 It took Jake 5 minutes to shear one sheep. How long would it take to shear:

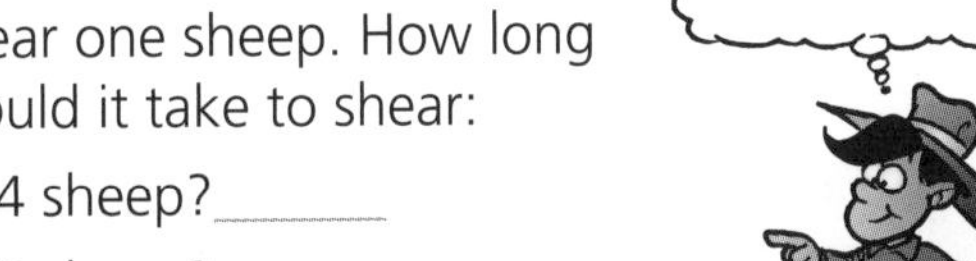

a 4 sheep? ______

b 8 sheep? ______

8 How many sheep could Jake shear in:

a 25 minutes? ______ b 50 minutes? ______

9 Circle the improper fraction. $\frac{4}{7}$ $\frac{5}{2}$ $2\frac{3}{5}$

10 What is the time?

a digital:

______ : ______

b analog:

______ past ______

11 What time is it 26 minutes after 3:58 pm? ______

12 What is the change from \$2 if you spend:

a 35 cents? ______ b 70 cents? ______

30:4

Extension

out of 4

1 If we choose one number from each column:

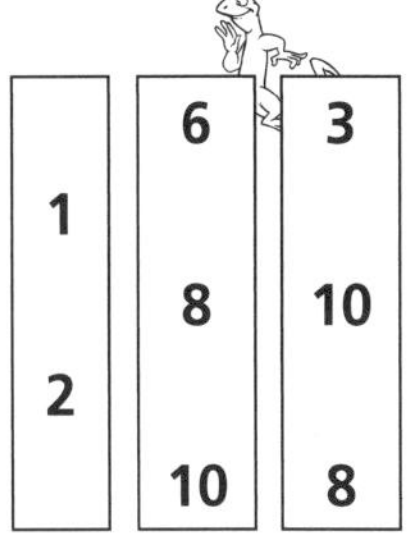

a which numbers give a total of 16? ______

b how many totals occur twice? ______

2 If three cars will fill a semi-trailer, how many semis can be filled using 21 cars? ______

3 4, 8, 12, 16, 20, ______

4 If this number pattern continued, what would be the:

a 8th number? ______

b 10th number? ______

Challenge

Follow these tracks, writing an answer in each circle.

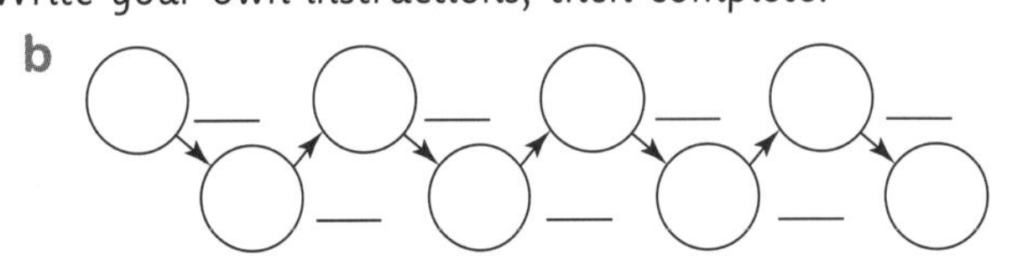

Write your own instructions, then complete.

b

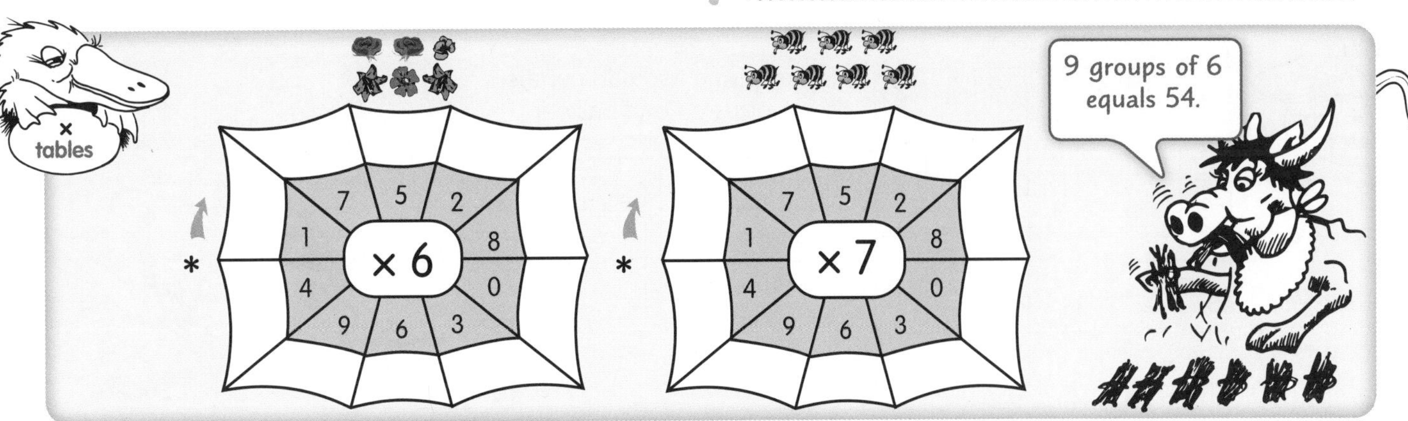

  • ISBN 978 0 6557 0884 1

31:1 [ ] out of 18

1. 3 × ______ = 15
2. 15 ÷ 3 ______
3. 10 × ______ = 30
4. 30 ÷ 10 ______
5. $\begin{array}{r} 350 \\ -\ 121 \\ \hline \end{array}$
6. 6 × ______ = 24
7. 24 ÷ 6 ______
8. 50 shared by 10 ______
9. 18 divided by 3 ______
10. $\begin{array}{r} 352 \\ +\ 423 \\ \hline \end{array}$

11. Write an equivalent fraction for:

    a $\frac{1}{2} = \frac{\square}{\square}$ **b** $\frac{1}{3} = \frac{\square}{\square}$

12. Circle: An odd number minus an even number equals an **odd** / **even** number.
13. **a** 17 − 8 = ______ **b** 15 − 6 = ______
14. What is the time?

    a digital: ______ : ______

    b analog: ______ past ______

15. I started a game of tennis at 3:13 pm and finished at 4 pm. For how long was I playing? ______
16. Share 20 cards among 5 players.

One share = ______ 20 ÷ 5 = ______

17. 5 groups of 5 = ______
18. 70 fingers. How many people? ______

31:2 [ ] out of 14

1. 8 × ______ = 48
2. 48 ÷ 8 ______
3. 8 × ______ = 72
4. 72 ÷ 8 ______
5. $\begin{array}{r} 400 \\ -\ 365 \\ \hline \end{array}$
6. 56 divided by 7 ______
7. 64 shared by 8 ______
8. Divide 81 by 9. ______
9. 45 in groups of 9 ______
10. $\begin{array}{r} 800 \\ -\ 398 \\ \hline \end{array}$
11. 卌 卌 卌 卌 II

    This tally stands for ______.

12. Circle: An even number plus an even number equals an **odd** / **even** number.
13.

July							August							September						
S	M	T	W	T	F	S	S	M	T	W	T	F	S	S	M	T	W	T	F	S
	1	2	3	4	5	6					1	2	3	1	2	3	4	5	6	7
7	8	9	10	11	12	13	4	5	6	7	8	9	10	8	9	10	11	12	13	14
14	15	16	17	18	19	20	11	12	13	14	15	16	17	15	16	17	18	19	20	21
21	22	23	24	25	26	27	18	19	20	21	22	23	24	22	23	24	25	26	27	28
28	29	30	31				25	26	27	28	29	30	31	29	30					

For the calendar above:

a How many full weeks in August? ______

b Write the date of the first Monday in September. ______

c What day is July 23? ______

d Days from July 23 to September 9? ______

14. Show each time on the clock.

    a 2:45 **b** 10:25

×	3	7	6	9	8	4
4						
6						
7						
8						
9						

Multiplying two odd numbers gives an odd answer.

Odd numbers end with 1, 3, 5, 7, 9.	Even numbers end with 2, 4, 6, 8.

These are odd numbers.

21, 27, 3, 49, 63, 81

31:3 ☐ out of 15

1. 3)24   2. 6)30   3. 8)32

4. 3)12   5. 4)32   6. 6)48

7. Circle: An odd number plus an even number equals an **odd** / **even** number.

8. The time half an hour after 4:35 is ________.

9. I baked a cake from 2:26 pm until 3 pm. For how long did it bake? ________

10. **a** The digital time is ______ : ______.

    b The analog time is ______ past ______.

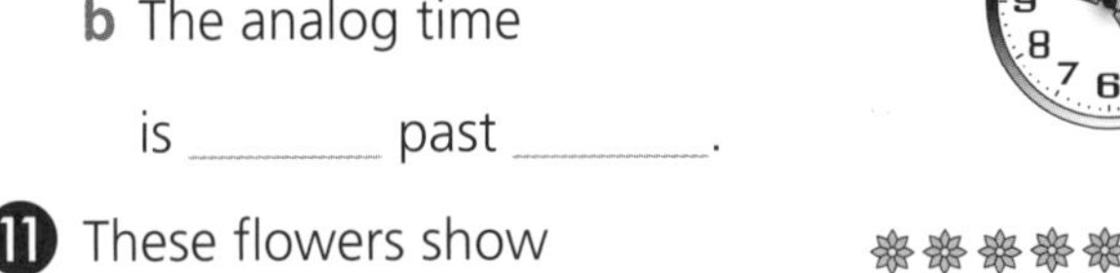

11. These flowers show ______ rows of ______ = ______.

    Share 42 flowers among 6 girls. One share = ______.

    $42 \div 6 =$ ______

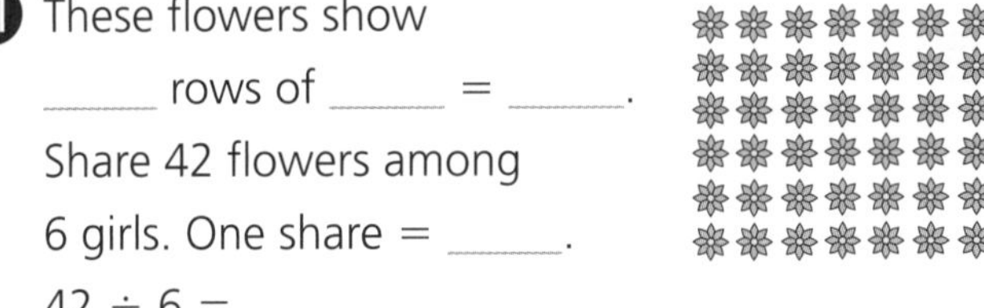

12. Write the numeral fifty-nine thousand, two hundred and forty-nine. ________

13. Write an equivalent fraction for:

    a $\frac{4}{4} = \frac{\square}{\square}$ **b** $\frac{3}{5} = \frac{\square}{\square}$

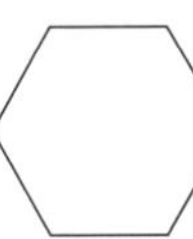

14. For this hexagon, how many:

    a sides? ______   **b** diagonals? ______

15. 61 − 46 ________

61

31:4 Extension ☐ out of 6

1. Patrick went fishing from 6:25 am until 8:19 am.

   For how long was he fishing?

   ____

2. 
   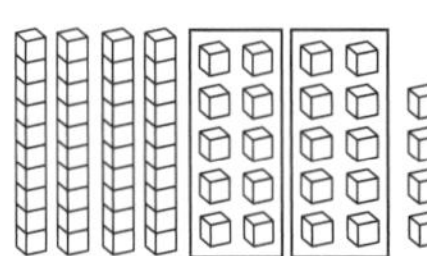

   Share $64 among 4 money boxes.

   One share = ________

3. Teagan has 33 birds. This is 3 times as many as Bill. How many does Bill have? ________

4. The Roman numeral for:

   a 8 ______   **b** 9 ______

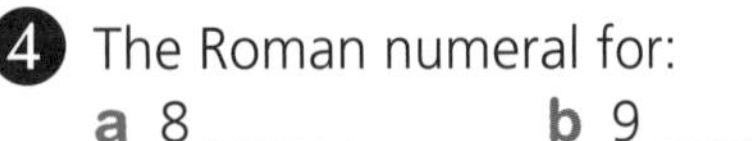

5. Rhonda has $176 and Alan has $124. How much must Rhonda give Alan if they are to have the same amount? ________

6. If # means 'add 4' and ~ means 'subtract 7', then:

   a 87 ##~~~ = ______   **b** 98 ##~~ = ______

Challenge

Draw and label two mixed numbers.

Concept

Division and repeated subtraction

How many groups of 4 shoes in 22?

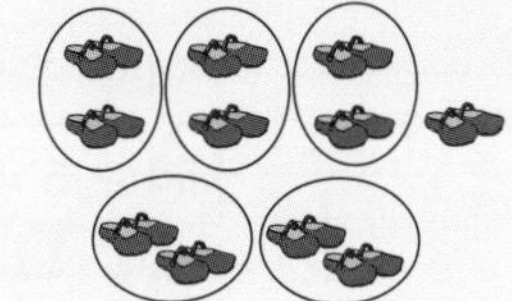

Answer:
5 groups, with 2 left.

There are 24 notes. How many:

a groups of 3? ______   **b** groups of 4? ______

c groups of 6? ______   **d** groups of 2? ______

e groups of 12? ______   **f** groups of 8? ______

g groups of 5? ______ with ____ left over.

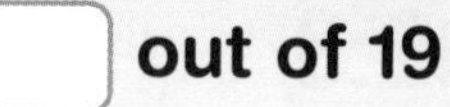

32:1 ☐ out of 18

1. 5 × 2 ____
2. 5 × 4 ____
3. 7 ×10 ____
4. 73 × 10 ____
5. 78 − 46
6. Double 6 × 2. ____
7. Double 5 × 4. ____
8. Double 4 × 3. ____
9. 8 × 3 ____
10. 453 + 238

11. How many minutes before 5 o'clock?

a  ____

b ____

12. 
These stars show ____ rows of ____ = ____
Share 15 stars between 3 girls.
One share = ____ 15 ÷ 3 = ____

13. If 3 × 5 = 15 does 4 × 5 equal 5 more? ____
14. If 5 × 8 = 40 then 8 × 5 = ____
15. True or false? To divide by 4 we can halve and halve again. ____
16. **a** 8 ÷ 4 = ____ **b** 12 ÷ 4 = ____
17. True of false? To multiply by 10 we can put a zero at the end of a whole number. ____
18. Is $\frac{6}{3}$ larger than $\frac{4}{3}$? ____

32:2 ☐ out of 19

1. 24 ÷ 4 ____
2. 44 ÷ 4 ____
3. 28 ÷ 4 ____
4. 40 ÷ 8 ____
5. 826 − 238
6. 200 times 6. ____
7. Multiply 6 by 30. ____
8. 40 multiplied by 7. ____
9. 500 times 7. ____
10. 946 − 389

11. What 4 coins could you use to make up 75 cents? ____
12. We share 30 cards equally and each of us has 6 cards. How many of us were there? ____
13. If 5 × 11 = 55 then 5 × 12 = ____
14. If 9 × 8 = 72 then 8 × 9 = ____
15. **a** If 2 × 3 = ____ then 3 × 3 = ____
b Does 3 × 3 equal 2 × 3 + 3? ____

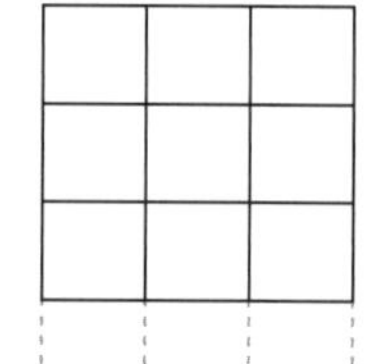

16. True or false? To divide by 8 we can halve the number 3 times. ____
17. **a** 80 ÷ 4 = ____ **b** 48 ÷ 8 = ____
18. Is 12 × 3 the same as (3 × 10) + (3 × 2)? ____

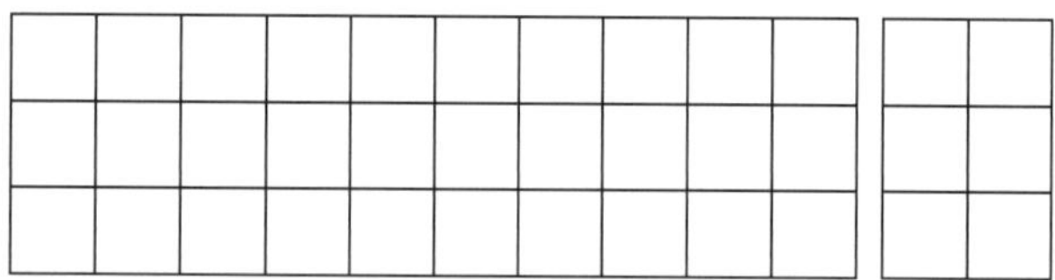

19. **a** How many grams in 6 kilograms? ____
b How many kg in 7000 g? ____
c How many kg in 400g? ____

Division linked with multiplication

If 7 × 9 = 63,
then 63 ÷ 7 = 9
63 ÷ 9 = 7.

a	7 × 6 = ☐	42 ÷ 7 = ____ 42 ÷ 6 = ____
b	4 × 7 = ☐	28 ÷ 4 = ____ 28 ÷ 7 = ____
c	6 × 8 = ☐	48 ÷ 6 = ____ 48 ÷ 8 = ____
d	9 × 6 = ☐	54 ÷ 9 = ____ 54 ÷ 6 = ____
e	8 × 9 = ☐	72 ÷ 8 = ____ 72 ÷ 9 = ____
f	8 × 7 = ☐	56 ÷ 8 = ____ 56 ÷ 7 = ____

 ISBN 978 0 6557 0884 1

32:3 out of 13

1. $\begin{array}{r} 718 \\ -\ 563 \\ \hline \end{array}$

2. $\begin{array}{r} 637 \\ 141 \\ +\ 242 \\ \hline \end{array}$

3. $\begin{array}{r} 321 \\ 434 \\ +\ 157 \\ \hline \end{array}$

4. $6\overline{)36}$

5. $2\overline{)14}$

6. $3\overline{)15}$

7. a These apples show ______ rows of ______ = ______

   b Share 18 apples between 3.
   One share = ______ 18 ÷ 3 = ______

8. If 5 × 7 = 35 then 7 × 5 = ______

9. a If 3 × 5 = ______
   then 4 × 5 = ______

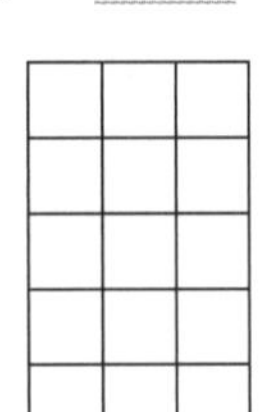

   b Does 4 × 5 equal
   3 × 5 + 5? ______

10. 88 people sat in groups of 8.
How many groups were there? ______

11. a 5 × 10 = ______ b 5 × 3 = ______

   c Combine these to find 5 times 13? ______

12. Estimate the capacity of a:
   a bucket. ______ b teaspoon. ______

13. a How many kilograms in 9000 grams? ______
   b How many kilograms in 4000 grams? ______

32:4 out of 7

Extension

1. I make 4 sandwiches from 2 slices of bread. How many can I make from 46 slices? ______

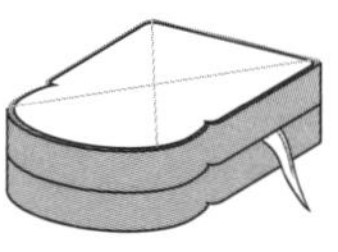

2. Use this diagram to find 6 × 23. ______

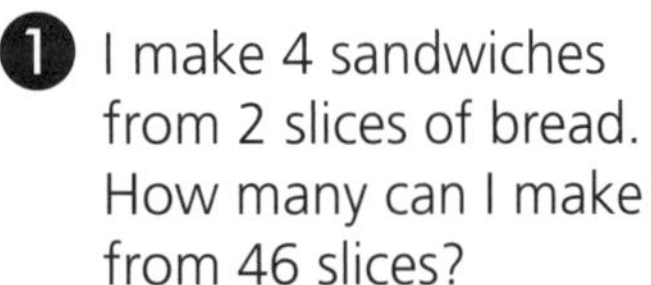

3. Find the area of the:

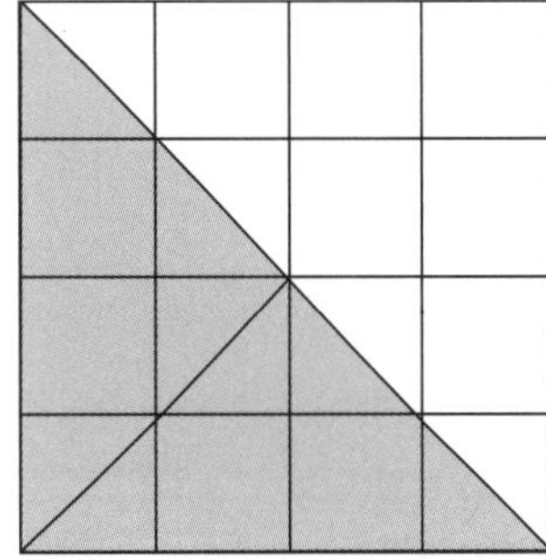

   a large square ______ cm^2
   b two triangles ______ cm^2
   c small triangle ______ cm^2

4. Would it be hot if the temperature was 35° Celsius? ______

5. What is the height of a tens block? ______

6. 800 + 34 000 + 900 000 + 23 ______

7. 824 ÷ 8 = ______

Challenge

Write related multiplication facts.
e.g. 2 × <u>4</u> = 8 so 2 × <u>8</u> = 16

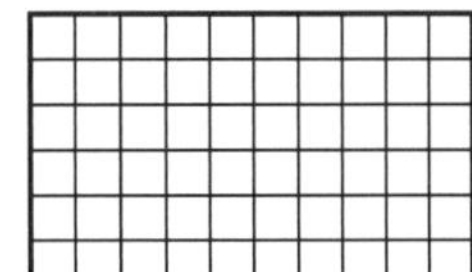

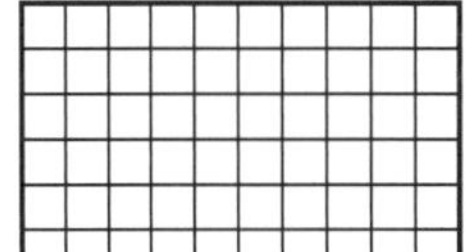

3 cm	12 m
$1\frac{1}{2}$ m	100 m
2 m	50 m
30 m	5 m
3 m	8 mm
30 cm	10 mm
1 mm	10 cm

Choose a measurement from the table as an estimate for the height of a:

a door ______ b giraffe ______

c cat ______ d mouse ______

e mug ______ f power pole ______

g bus ______ h side fence ______

33:1 out of 18

1. 6 × 2 ____
2. 6 × 4 ____
3. 10 × 2 ____
4. 20 × 2 ____
5. $\begin{array}{r} 87 \\ +58 \\ \hline \end{array}$
6. 5 × 3 + 5 ____
7. 7 × 2 + 7 ____
8. Double 5 × 3. ____
9. Double 7 × 3. ____
10. $\begin{array}{r} 857 \\ -279 \\ \hline \end{array}$
11. If 3 × 9 = 27 then 9 × 3 = ____
12. **a** If 2 × 6 = ____

    then 3 × 6 = ____

    b Does 3 × 6 equal 2 × 6 + 6? ____

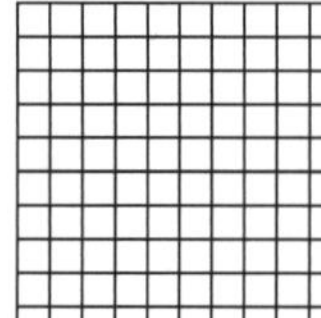

13. Colour 5 tenths of this hundreds block and write the number as a decimal.

    ____ . ____

14. Is 0·5 larger than 0·05? ____
15. Estimate how many maths textbooks would be needed to cover one square metre. ____
16. Write the short form for:

    a 6 kilograms ____ **b** 46 grams ____
17. 6 rows of 4 children = ____ children
18. **a** ____ × 9 = 9 **b** ____ × 3 = 12

    c 9 ÷ 9 = ____ **d** 12 ÷ 3 = ____

33:2 out of 16

1. 60 ÷ 4 ____
2. 64 ÷ 8 ____
3. 80 ÷ 4 ____
4. 80 ÷ 8 ____
5. $\begin{array}{r} 564 \\ +278 \\ \hline \end{array}$
6. Double 7 × 8. ____
7. 14 × 8 ____
8. Double 7 × 9. ____
9. 14 × 9 ____
10. $\begin{array}{r} 693 \\ -387 \\ \hline \end{array}$
11. 4 × ____ = 36 so 36 ÷ 4 = ____
12. Three girls shared 15 cards. How many did each girl get? ____
13. If 4 × 8 = 32 then 8 × 4 = ____
14. Explain how you could find 3 × 27 = ____

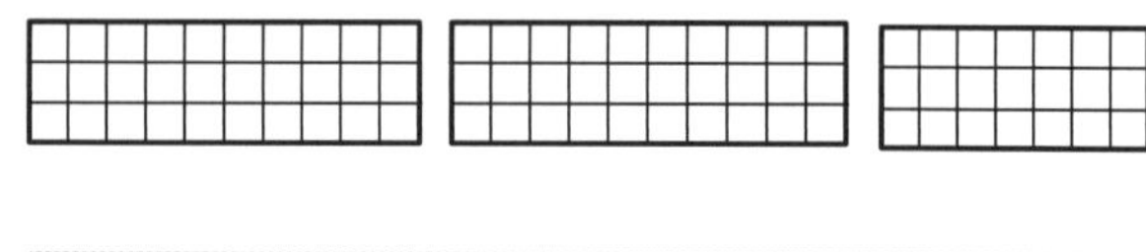

15. Estimate (E) then find the area (A) of this rectangle.

    E = ____ cm^2

    A = ____ cm^2

    What is the area of each triangle?

    A = ____

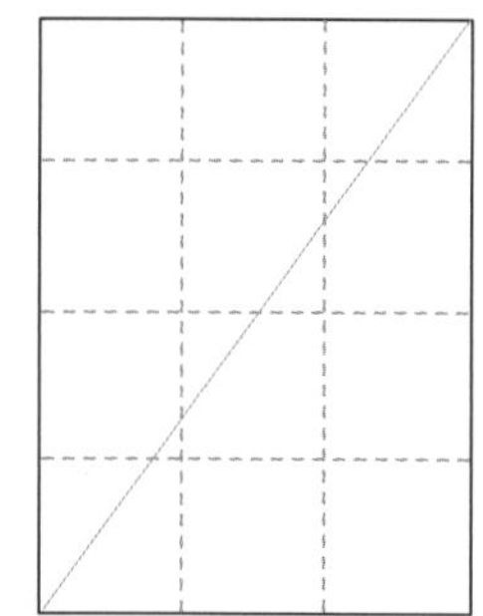

16. If the temperature is 8°C would it be hot or cold? ____

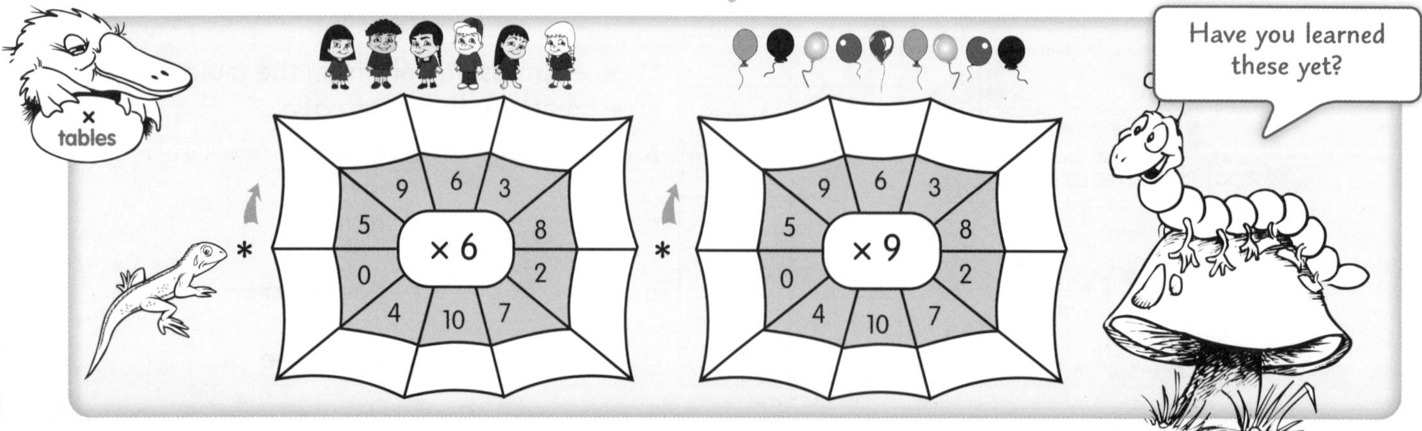

 • *AUSTRALIAN SIGNPOST MATHS 4 MENTALS* • ISBN 978 0 6557 0884 1

33:3 out of 13

1. $\begin{array}{r} 629 \\ 231 \\ +117 \\ \hline \end{array}$

2. $\begin{array}{r} 800 \\ -298 \\ \hline \end{array}$

3. $\begin{array}{r} 564 \\ -159 \\ \hline \end{array}$

4. $2\overline{)16}$

5. $2\overline{)4}$

6. $4\overline{)16}$

7. If 6 × 7 = 42 then 7 × 6 = ______

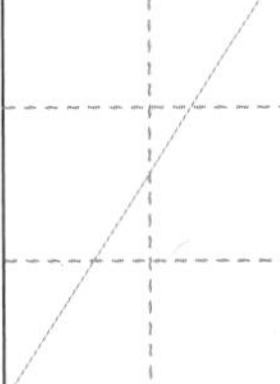

8. **a** The area of this rectangle is ______ cm².

   b The area of each triangle is ______ cm².

9. **a** How many kilograms in 2000 grams? ______

   b How many kilograms in 6000 grams? ______

10.   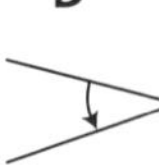

Put these angles in order with the smallest first. ______

11. Write these measurements using one decimal place.

    a 700 g = ______ kg **b** 400 g = ______ kg

12. How could you put 24 people into equal groups?

______

______

13. How many triangles like this are needed to cover the second shape? ______

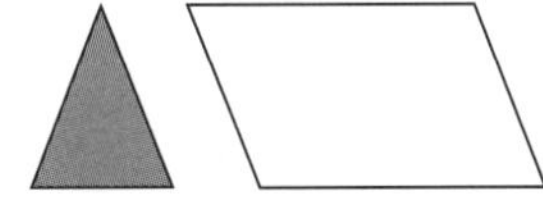

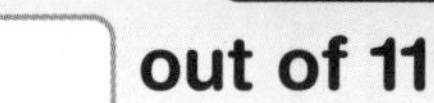

33:4 out of 11

1. If 365 × 18 = 6570 then 18 × 365 = ______
2. Use this diagram to find 9 × 63. ______

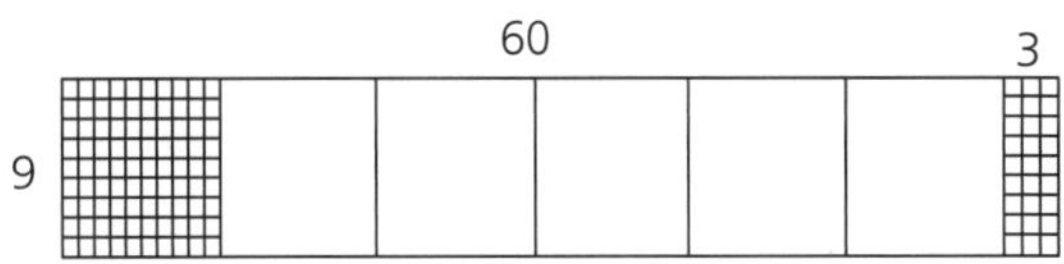

3. What is the mass of 1 L of water? ______
4. True or false: Two of your steps is close to 1 m in length. ______
5. If □ − 8 = 4, and 30 − △= 26, find the answer to: □ ÷ △. ______
6. If 36 × 11 = 396, what does 35 × 11 equal? ______
7. **a** 5 × 3 × 30 ______ **b** 6 × 5 × 40 ______
8. **a** 3 × 3 × 3 ______ **b** 3 × 3 × 3 × 3 ______
9. How many 5 cent coins have the same value as $2? ______
10. How many sides on 24 triangles? ______
11. If 14 × 14 = 196, what does 15 × 14 equal? ______

Challenge

Use different colours to make shapes with an area of 5 cm².

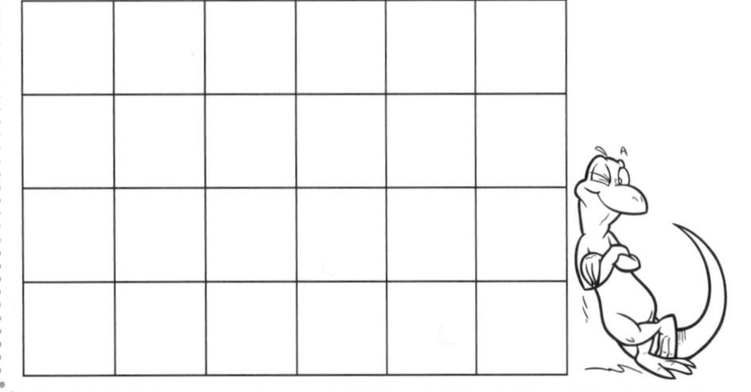

Turn to ID card B on page 7.
Give the answers for these numbers.

(11) ______ angle (12) ______ angle

(13) ______ angle (14) ______ angle

(15) ______ angle (16) ______

(17) ______ of an angle (18) ______ of an angle

The angle at a square corner is called a right angle.

34:1 ☐ out of 16

1. 145 + 7 ______
2. 100 − 9 ______
3. 8 × 5 ______
4. 40 ÷ 8 ______
5. $\begin{array}{r} 85 \\ -49 \\ \hline \end{array}$
6. 3·3, 3·4, ______, ______
7. \$5 + \$2 + 50c ______
8. \$5 + \$5 + 20c ______
9. \$20 + 50c + 20c ______
10. $\begin{array}{r} 500 \\ +393 \\ \hline \end{array}$
11. Write the short form for:
    a 28 grams ______
    b 13 kilograms ______
12. The number of these rectangles that would be needed to cover:

     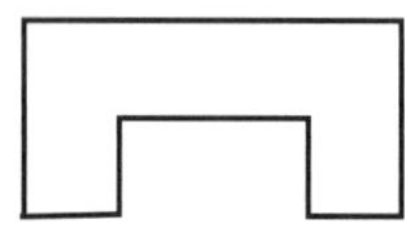
    a ______ b ______
13. a ______ × 5 = 35 so 35 ÷ 5 = ______
    b ______ × 10 = 80 so 80 ÷ 10 = ______
14. Draw all the axes of symmetry.
    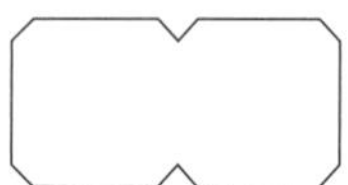
15. Does spinner A or B have the best chance of spinning grey?
    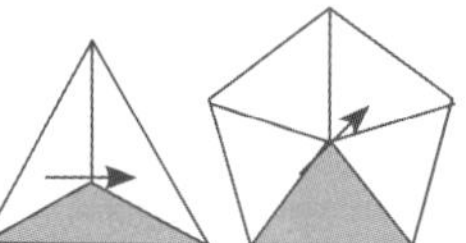
    ______
16. List the value of the notes and coins needed to make \$7.20. ______

34:2 ☐ out of 15

1. 153 + 249 ______
2. 320 + 190 ______
3. 15 × 4 ______
4. 100 ÷ 2 ______
5. $\begin{array}{r} 600 \\ -379 \\ \hline \end{array}$
6. 9·1, 9·2, ______, ______
7. \$50 + \$20 + 50c ______
8. \$20 + 50c + 50c ______
9. \$5 − \$1.45 ______
10. $\begin{array}{r} 579 \\ +362 \\ \hline \end{array}$
11. ______ × 8 = 64 so 64 ÷ 8 = ______
12. Explain how you could use this diagram to find 7 × 22 = ______
    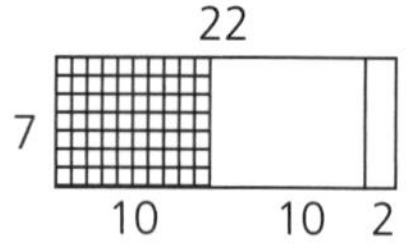

    ______
13. 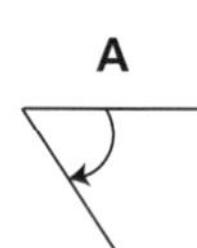

    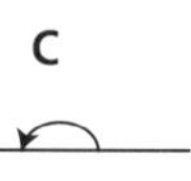

    Which angle is:
    a a straight angle? ______
    b an acute angle? ______
14. My change from \$10 if I spend \$3.55 ______
15. I need to put 32 people into equal groups. How could this be done? ______

÷ tables

18	9	27		
0	÷ 3	3		
12		15		
24	6	21		

30	6	54
12	÷ 6	42
36		24
18	48	60

How many groups of 3?
How many rows of 6?

34:3 out of 10

1.
$$\begin{array}{r} 629 \\ 231 \\ +117 \\ \hline \end{array}$$

2.
$$\begin{array}{r} 800 \\ -298 \\ \hline \end{array}$$

3.
$$\begin{array}{r} 564 \\ -159 \\ \hline \end{array}$$

4. ____ × 9 = 72 so 72 ÷ 9 = ____

5. Explain how you could use this diagram to find 8 × 47 = ____

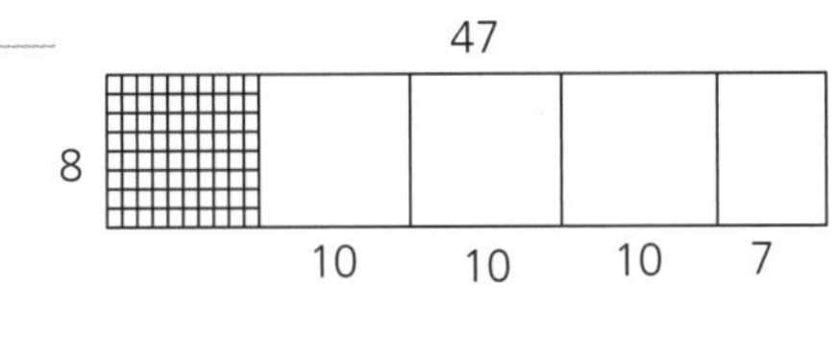

______

______

______

6. Circle the shape that has been cut out.

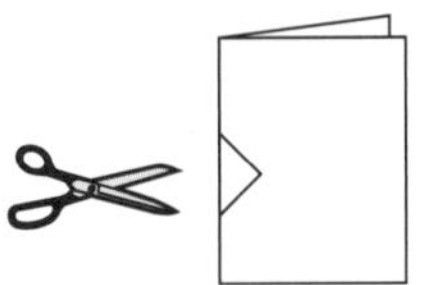

A  B  C 

7. Write the short form for:
   - a 378 grams ______
   - b 92 kilograms ______

8. a ____ × 8 = 56 so 56 ÷ 8 = ____

   b ____ × 7 = 49 so 49 ÷ 7 = ____

9. List the least number of notes and coins needed to make $15.85.

______

10. What is the change from $50 when I spend $27.50? ______

34:4 out of 6

Extension

1. Use this diagram to find 8 × 75. ______

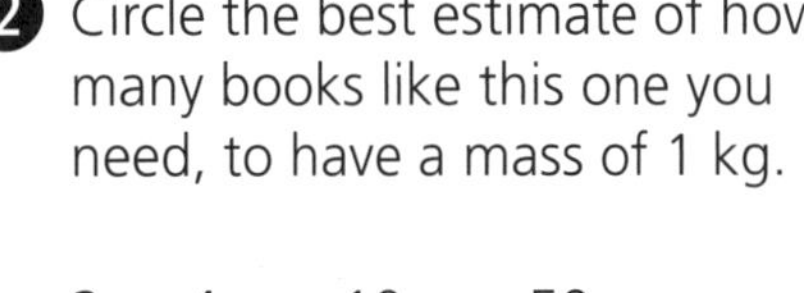

2. Circle the best estimate of how many books like this one you need, to have a mass of 1 kg.

2 4 10 50

3. Eight friends gave either $4 or $5 each. If $38 was given, what is the most number that could have given $5? ______

4. This figure has squares of different size. How many squares altogether? ______

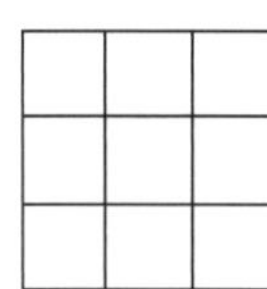

5. What is the change from $100 if I spend $83.45? ______

6. If 197 + 803 = 1000, then:
   - a 1000 − 803 = ____
   - b 1000 − 197 = ____

Challenge

What do you know about the number 49?

______

______

______

______

______

______

To round off correct to the nearest 5 cents, give the closest answer that ends with 5 or 0.

- a Is 52 cents closer to 50 cents or 55 cents?
- b Is 97 cents closer to 95 cents or $1?

Round off each of these correct to the nearest 5 cents.

c 63c ______	d 21c ______	e 88c ______	f 99c ______
g 72c ______	h 93c ______	i 26c ______	j 54c ______

The Royal Australian Mint stopped making 1c and 2c coins in 1992.

 • *AUSTRALIAN SIGNPOST MATHS 4 MENTALS* • ISBN 978 0 6557 0884 1

35:1 ☐ out of 16

1. 5 × 10 ______
2. 9 × 10 ______
3. 80 ÷ 10 ______
4. 170 ÷ 10 ______
5. $\begin{array}{r} 85 \\ -\,68 \\ \hline \end{array}$
6. Multiply 5 by 100. ______
7. 30 divided by 10 ______
8. 90 shared by 10 ______
9. 6 times 1000 ______
10. $\begin{array}{r} 469 \\ +\,283 \\ \hline \end{array}$
11. What 4 coins could you use to make up $2.25 cents? ______
12. Count on from $6.70 to find the change given for $10. ______

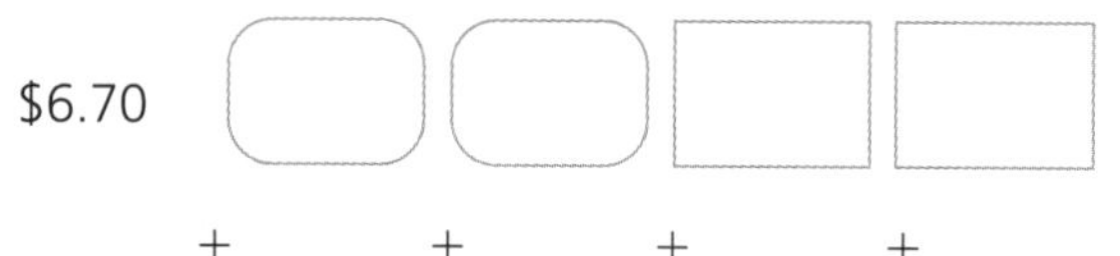

+ ______ + ______ + ______ + ______

13. Who is holding the horizontal line? ______

Emma Jon

14. Does a regular pentagon have:
    a symmetry? ______
    b rotational symmetry? ______

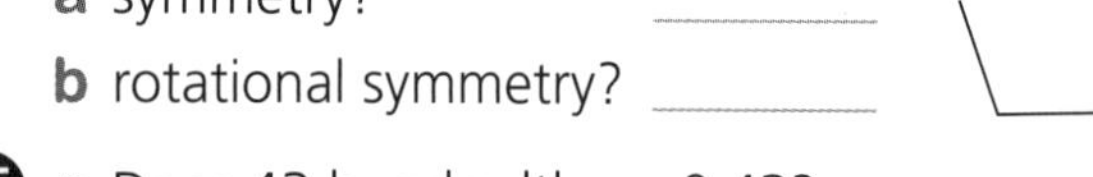

15. a Does 43 hundredths = 0·43? ______
    b Does 5 hundredths = 0·5? ______
16. a Which is larger, 0·5 or 0·36? ______
    b Which is larger, 0·46 or 0·6? ______

35:2

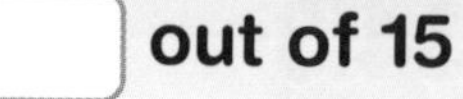

1. 5 × 60 ______
2. 100 × 8 ______
3. 40 ÷ 10 ______
4. 600 ÷ 10 ______
5. $\begin{array}{r} 500 \\ -\,375 \\ \hline \end{array}$
6. 940 × 10 ______
7. 350 × 10 ______
8. 53 × 100 ______
9. 39 × 1000 ______
10. $\begin{array}{r} 700 \\ -\,426 \\ \hline \end{array}$
11. Colour $4\frac{1}{4}$ squares.

12.

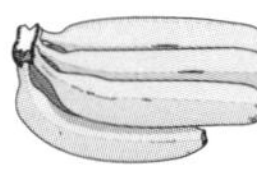

$5.07 $4.54 $5.61

a What is the total cost of the fruit? ______
b Round the total to the nearest 5c. ______

13. Count on from $35.90 to find the change given for $50. ______

$35.90

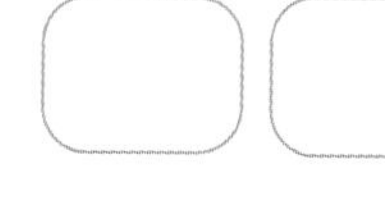

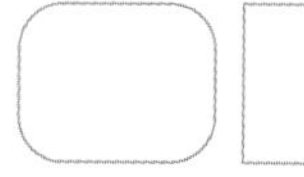

+ ______ + ______ + ______ + ______

14. Circle the shape that can tessellate.

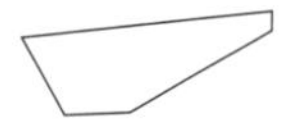

15. a 600 000 ÷ 10 ______ b 350 ÷ 10 ______
    c 300 000 ÷ 100 ______ d 260 × 10 ______

Crossnumber puzzle

Across

1 385 − 80
5 200 + 300
6 8159 − 10
8 4 × 6
9 761 − 50
10 2012 + 6

Down

1 360 − 12
2 5224 + 20
3 3 × 5
4 7040 + 1
7 1298 − 6
9 700 + 8

35:3 ☐ out of 12

1. 3)24
2. 6)30
3. 8)32
4. 3)12
5. 4)32
6. 6)48
7. Count on from \$24.50 to find the change given for \$50. ______

   \$24.50 ☐ ☐ ☐

   \+ ____ + ____ + ____
8. Complete this tessellation.
9. Does a regular hexagon have:

   a symmetry? ______

   b rotational symmetry? ______

10. **a** 20 000 ÷ 10 ______ **b** 490 ÷ 10 ______

    c 900 000 ÷ 100 ______ **d** 191 × 10 ______
11. Each pencil case holds 100 pencils. How many pencils could you fit into 63 pencil cases? ______
12. Find the total then round to the nearest 5c.

    a apples (\$4.36), bananas (\$5.36) and strawberries (\$7.55) ______

    b carrots (\$3.55), onions (\$5.37) and celery (\$3.87) ______

35:4 Extension ☐ out of 7

1. If this pattern continues, how many hexagons would be in row:

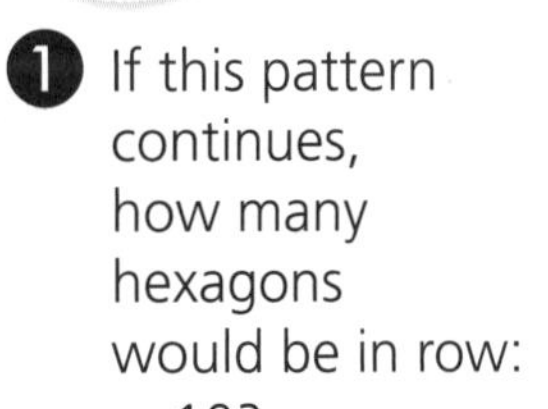

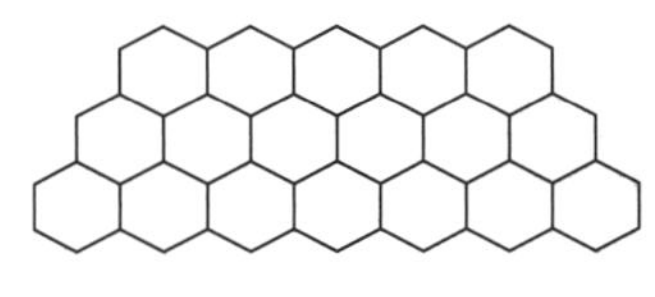

   a 10? ______ **b** 100? ______
2. List regular 2D shapes that can tessellate.

   ______

   ______
3. How many different ways are there of choosing two of these squares side by side for a handball court?

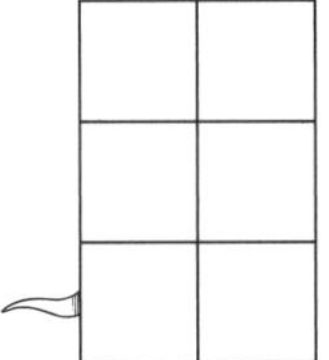

   ______
4. **a** 857 630 ÷ 10 ______ **b** 920 ÷ 10 ______

   c 745 300 ÷ 100 ______ **d** 576 × 10 ______
5. I have 24 ten-dollar notes. How much money do I have? ______
6. If 14 × 14 = 196, what does 15 × 14 equal? ______
7. \$200 − \$123.60 ______

Challenge

Ask at least 4 people what time they went to bed last night. Record their answers and compare your results with your classmates' results.

before 8 pm	
between 8 pm and 9 pm	
between 9 pm and 10 pm	
between 10 pm and 11 pm	
after 11 pm	

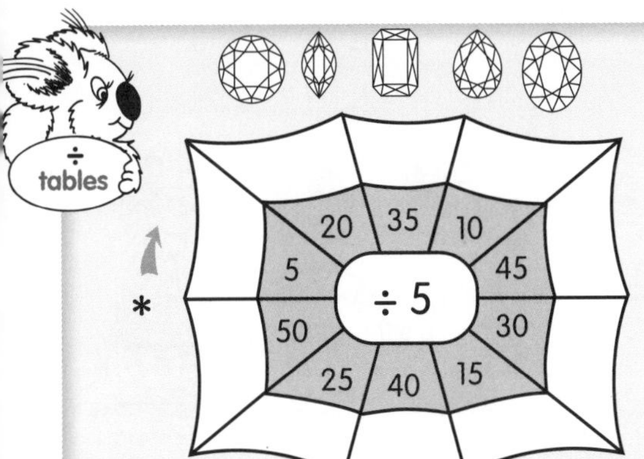

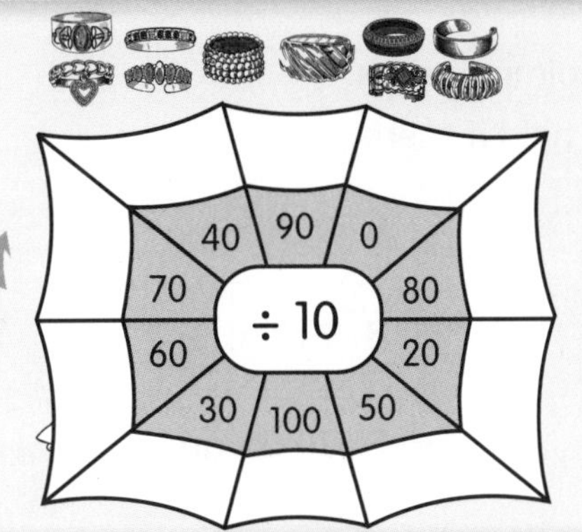

36:1

out of 18

1. 6 × 10 ______
2. 9 × 100 ______
3. 3 × 100 ______
4. 8 × 10 ______
5. $\begin{array}{r} 56 \\ -37 \\ \hline \end{array}$
6. 90 divided by 10 ______
7. 40 shared by 10 ______
8. Divide 300 by 10. ______
9. 700 ÷ 100 ______
10. $\begin{array}{r} 56 \\ -48 \\ \hline \end{array}$
11. Which pipes are:
    a horizontal? ______
    b vertical? ______
    c sloping? ______
    d not horizontal? ______

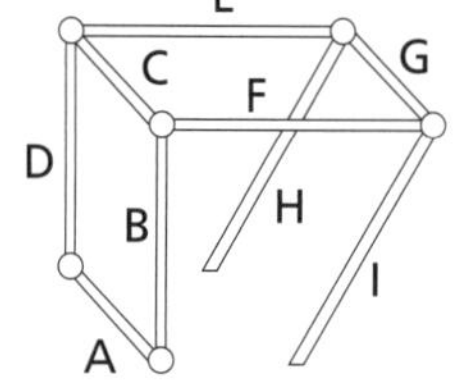

12. a ____ × 3 = 15 so 15 ÷ 3 = ____
    b ____ × 10 = 40 so 400 ÷ 10 = ____
13. How many sides on 5 octagons? ______
14. ____ × 7 = 70
15. Multiply each number by 100.
    a 8 ______ b 5 ______ c 2 ______
16. Legs on:
    a 10 crocodiles. ______
    b 5 crocodiles. ______
    c 8 crocodiles. ______

17. Colour $4\frac{1}{2}$ of these shapes.

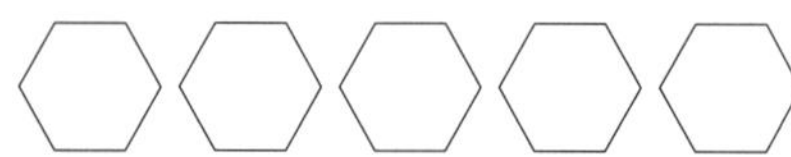

18. How many sides on 6 pentagons? ______

36:2

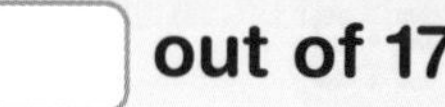

out of 17

1. 3647 × 10 ______
2. 8402 × 10 ______
3. 435 × 100 ______
4. 894 × 100 ______
5. $\begin{array}{r} 563 \\ +592 \\ \hline \end{array}$
6. 84 630 ÷ 10 ______
7. 78 300 ÷ 10 ______
8. 63 800 ÷ 100 ______
9. 200 × 1000 ______
10. $\begin{array}{r} 730 \\ -395 \\ \hline \end{array}$

11. A B C

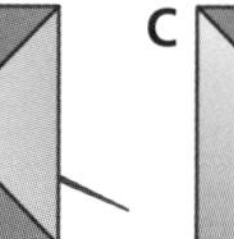

    a Which spinner has an even chance of spinning grey? ______
    b Which spinner has an unlikely chance of spinning grey? ______
12. Circle the shape that can tessellate.

13. a 67 000 ÷ 10 ______ b 880 ÷ 10 ______
    c 500 000 ÷ 100 ______ d 250 × 10 ______
14. Does a square have rotational symmetry? ______
15. 1260 students at our school paid \$10 for the disco fundraiser. How much money was paid altogether? \$______
16. If 15 × 26 = 390, then:
    a 390 ÷ 15 = ______ b 390 ÷ 26 = ______
17. a 25 − 8 = 30 − ______ b 8 × ______ = 40
    c 37 − 6 = 40 − ______ d 3 × ______ = 27
    e 45 − 3 = 50 − ______ f 4 × ______ = 20

Friends are in a line holding hands.
How many hands are held if there are:

a 2 children? ______ b 3 children? ______

c 4 children? ______ d 7 children? ______

e 10 children? ______ f 20 children? ______

36:3   ☐ out of 7

1

H	T	U
6	0	3
− 2	7	9

2

H	T	U
5	2	4
− 1	3	6

3 10 identical lamps cost $3500 in total. How much did each lamp cost? ______

4 If 13 × 27 = 351, then:

a 351 ÷ 13 = ______   b 351 ÷ 27 = ______

5 a 453 + 97 = 453 + 100 − ______ = ______

b 637 + 95 = 637 + 100 − ______ = ______

c 254 + 92 = 254 + 100 − ______ = ______

6 5400 mm = ______ m

7 

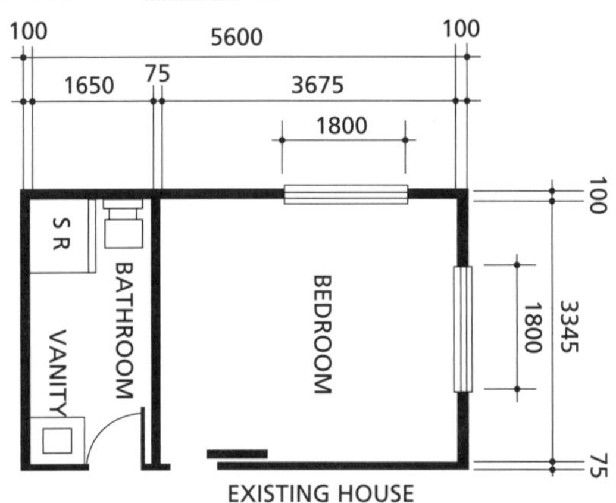

The lengths on this map are written in millimetres. What is the length of the:

a longer side of the bedroom?

______ mm or ______ m ______ mm

b shorter side of the bathroom?

______ mm or ______ m ______ mm

36:4   ☐ out of 7

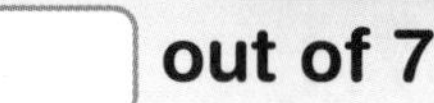

1 What was the year 100 years before 2024? ______

2 True (T) or false (F)?

a 19 × 12 = 12 × 19 ______

b 19 + 12 = 12 + 19 ______

3 The Roman numeral for 10. ______

4 Does each row, column and diagonal of this magic square, add up to 18?

7	8	3
2	6	10
9	4	5

______

5 In how many ways can you make 35 cents using 5c, 10c, and 20c coins? ______

6 a

$$\begin{array}{r} 80 \\ - \square \\ \hline 24 \\ \hline \end{array}$$

b

$$\begin{array}{r} \square \\ - 32 \\ \hline 17 \\ \hline \end{array}$$

7 If # means 'add 3' and ~ means 'subtract 2', then 8 # # # ~ ~ ~ = ______

Challenge

Create and answer your own questions using the code in question 7 above.

______

______

______

______

______

______

______

Rounding to the nearest 5 cents

To round to the nearest 5 cents, give the closest answer that ends in 5 or 0.

a Is $1.37 closer to $1.35 or $1.40? ______

b Is $1.32 closer to $1.30 or $1.35? ______

Round each of these to the nearest 5 cents.

c $2.48 ______   d $8.13 ______   e $6.01 ______   f $1.99 ______

37:1 out of 18

1. 7 − ____ = 3
2. ____ − 4 = 7
3. 14 − ____ = 8
4. ____ + 9 = 17
5. $\begin{array}{r} 57 \\ -38 \\ \hline \end{array}$
6. 17 + ____ = 36
7. ____ − 5 = 15
8. 10 + ____ = 13 + 5
9. 25 ÷ 5 = ____
10. $\begin{array}{r} 643 \\ +285 \\ \hline \end{array}$
11. If 8 × 6 = 48, then:
    a 48 ÷ 6 = ____ b 48 ÷ 8 = ____
12. Bridge to 10 to find:
    a 18 + 7 = ____ b 27 + 8 = ____
13. a 37 − 19 = ____ − 20
    b 35 − 18 = ____ − 20
14. Use the scale to find how far the emu is from the meeting place. ____

Scale: 1 cm = 1 km ——

Meeting place 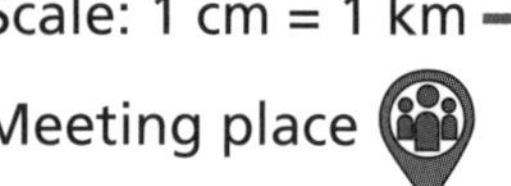 Emu 

15. Surveys are used to collect ________.
16. a 45 + 27 = ____ + 30
    b 53 + 38 = ____ + 40
17. What is your age now?
    ____ years and ____ months
18. How many tenths are in 5·6? ____

37:2 out of 14

1. 6 × ____ = 42
2. ____ ÷ 5 = 6
3. 7 + ____ = 18
4. 5 × ____ = 35
5. $\begin{array}{r} 56 \\ -38 \\ \hline \end{array}$
6. 35 ÷ 7 + ____ = 12
7. 30 ÷ 5 + ____ = 15
8. 50 ÷ 5 + ____ = 18
9. 18 + ____ = 26 + 4
10. $\begin{array}{r} 369 \\ +573 \\ \hline \end{array}$
11. If 14 × 32 = 448 then:
    a 448 ÷ 14 = ____ b 448 ÷ 32 = ____
12. Partition to complete the diagram and find:
    a 56 + 38 = ____ b 67 − 48 = ____

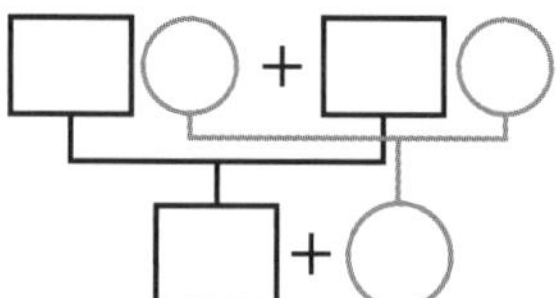

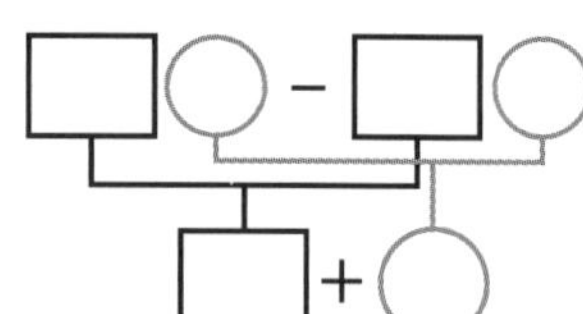

13.

A	B 6 x 4	C 3 x 4
D 3 x 3	E 4 x 6	F
G 6 lots of 4	H 3 lots of 3	I 6 x 4

a Is D equal to E? ____
b Is B equal to E? ____
Which of the above are equal to:
c 3 rows of 3? ____
d 6 groups of 4? ____

14. What is the change from $50 if I spend $5.30. ____

Turn to ID card A on page 6.
Give the answers for these numbers.

(1) ________ (2) ________
(3) ________ (4) ________
(5) ________ (6) ________
(7) ________ (8) ________

37:3 out of 6

1

H	T	U
6	0	3
− 2	7	9

2

H	T	U
5	2	4
− 1	3	6

3 a 65 − 28 = ______ − 30
b 44 − 27 = ______ − 30

4 a 38 − 5 = 40 − ______ b 7 × ______ = 49
c 49 − 7 = 50 − ______ d 6 × ______ = 36

5

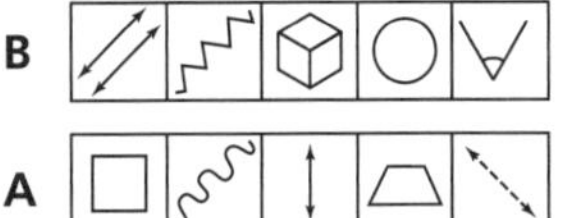

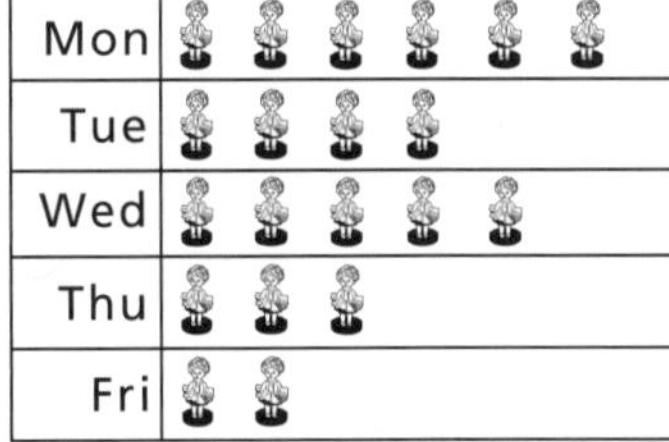

B, A / 1 2 3 4 5

What shape has coordinates:
a 3B? ____ b 1A? ____
Write the coordinates for the:
c trapezium ______
d parallel lines ______

6 **Dolls sold**

Mon	6 dolls symbols
Tue	4 doll symbols
Wed	5 doll symbols
Thu	3 doll symbols
Fri	2 doll symbols

(doll symbol) stands for 5 dolls

a On which day was the greatest number of dolls sold? ______
b How many dolls were sold on Wednesday? ______
c What was the total number of dolls sold? ______

37:4 Extension out of 6

1 Use place value to find:
a 586 − 139
586 − 100 = ______
486 − 30 = ______
456 − 9 = ______
b 723 − 218
723 − 200 = ______
523 − 10 = ______
513 − 8 = ______

2 If 15 × 16 = 240 then:
a 240 ÷ 15 = ______ b 240 ÷ 16 = ______

3 Partition to complete the diagram and find:
a 58 + 48 = ______ b 47 − 36 = ______
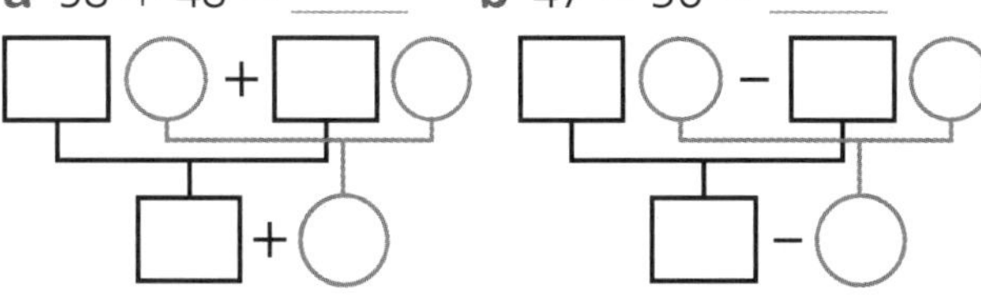

4 a 586 + 99 = 586 + 100 − ______ = ______
b 293 + 94 = 293 + 100 − ______ = ______

5 This figure has squares of difference size. How many squares altogether? ______
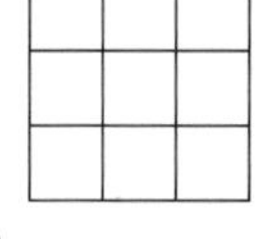

6 6 × ☐ + 4 = 52 ☐ = ______

Challenge
What do you know about the decimal 261·73?

______

Measure
Fill out this table about yourself, a relative or a friend.
Compare your answers to those on page 11 in Unit 1.

Name: ______ **Date:** ______

Age: ______	Mass: ______ kg	Shoe size: ______
Height: ______ cm	Waist: ______ cm	Neck size: ______ cm

 AUSTRALIAN SIGNPOST MATHS 4 MENTALS • ISBN 978 0 6557 0884 1

Tables of number and measurement

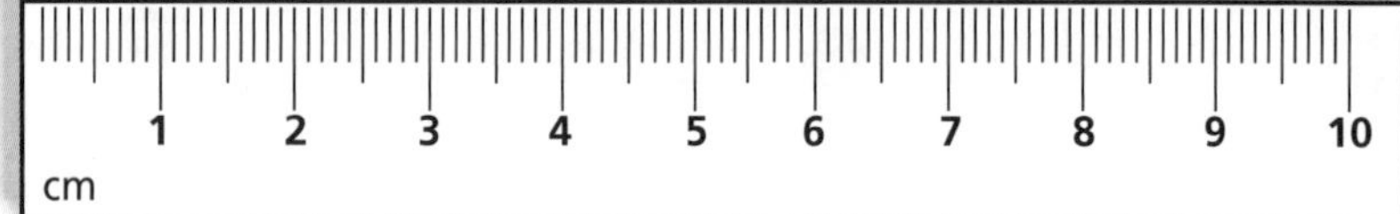

Length

1 centimetre = 10 millimetres
1 metre = 100 centimetres
1 metre = 1000 millimetres
1 kilometre = 1000 metres

The **freezing point** of water is **0°C**.
The **boiling point** of water is **100°C**.

°C means degrees Celsius.

A temperature of **5°C** is a **cold** day.
A temperature of **35°C** is a **hot** day.

Area

1 hectare = 10 000 m^2
1 square metre = 10 000 cm^2

Mass

1 kilogram = 1000 grams
1 tonne = 1000 kilograms

Roman Numerals

1	= I	**6**	= VI	**20**	= XX	**90**	= XC
2	= II	**7**	= VII	**30**	= XXX	**100**	= C
3	= III	**8**	= VIII	**40**	= XL	**200**	= CC
4	= IV	**9**	= IX	**50**	= L	**500**	= D
5	= V	**10**	= X	**60**	= LX	**1 000**	= M

Capacity and volume

1000 millilitres = 1 litre
1000 litres = 1 kilolitre
1 millilitre = 1 cm^3
1 litre = 1000 cm^3

5 mL

Months of the Year

Thirty days has September, April, June and November. All the rest have thirty-one, except February alone, which has twenty-eight days clear and twenty-nine days each leap year.

You can use the knuckles of your hands to find the number of days in each month.

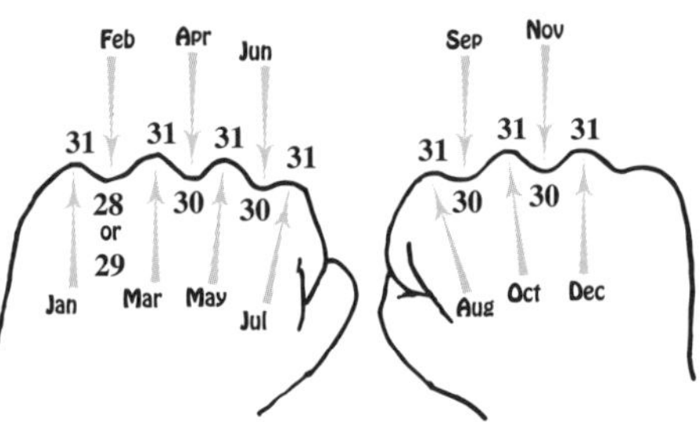

Time

1 minute = 60 seconds
1 hour = 60 minutes
1 day = 24 hours
1 week = 7 days
1 fortnight = 2 weeks
1 year = 52 weeks
1 year = 365 days
1 leap year = 366 days
1 decade = 10 years
1 century = 100 years

am stands for **ante meridiem**.
am means **before midday**.

pm stands for **post meridiem**.
pm means **after midday**.

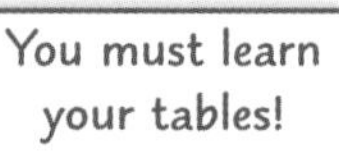

Seasons

Summer: December, January, February
Autumn: March, April, May
Winter: June, July, August
Spring: September, October, November

Multiplication Tables

1 × 2 = 2	1 × 3 = 3	1 × 4 = 4	1 × 5 = 5	1 × 6 = 6	1 × 7 = 7	1 × 8 = 8	1 × 9 = 9	1 × 10 = 10
2 × 2 = 4	2 × 3 = 6	2 × 4 = 8	2 × 5 = 10	2 × 6 = 12	2 × 7 = 14	2 × 8 = 16	2 × 9 = 18	2 × 10 = 20
3 × 2 = 6	3 × 3 = 9	3 × 4 = 12	3 × 5 = 15	3 × 6 = 18	3 × 7 = 21	3 × 8 = 24	3 × 9 = 27	3 × 10 = 30
4 × 2 = 8	4 × 3 = 12	4 × 4 = 16	4 × 5 = 20	4 × 6 = 24	4 × 7 = 28	4 × 8 = 32	4 × 9 = 36	4 × 10 = 40
5 × 2 = 10	5 × 3 = 15	5 × 4 = 20	5 × 5 = 25	5 × 6 = 30	5 × 7 = 35	5 × 8 = 40	5 × 9 = 45	5 × 10 = 50
6 × 2 = 12	6 × 3 = 18	6 × 4 = 24	6 × 5 = 30	6 × 6 = 36	6 × 7 = 42	6 × 8 = 48	6 × 9 = 54	6 × 10 = 60
7 × 2 = 14	7 × 3 = 21	7 × 4 = 28	7 × 5 = 35	7 × 6 = 42	7 × 7 = 49	7 × 8 = 56	7 × 9 = 63	7 × 10 = 70
8 × 2 = 16	8 × 3 = 24	8 × 4 = 32	8 × 5 = 40	8 × 6 = 48	8 × 7 = 56	8 × 8 = 64	8 × 9 = 72	8 × 10 = 80
9 × 2 = 18	9 × 3 = 27	9 × 4 = 36	9 × 5 = 45	9 × 6 = 54	9 × 7 = 63	9 × 8 = 72	9 × 9 = 81	9 × 10 = 90
10 × 2 = 20	10 × 3 = 30	10 × 4 = 40	10 × 5 = 50	10 × 6 = 60	10 × 7 = 70	10 × 8 = 80	10 × 9 = 90	10 × 10 = 100